Advances in Intelligent Systems and Computing

Volume 273

For further volumes:
http://www.springer.com/series/11156

About this Series

The series "Advances in Intelligent Systems and Computing" contains publications on theory, applications, and design methods of Intelligent Systems and Intelligent Computing. Virtually all disciplines such as engineering, natural sciences, computer and information science, ICT, economics, business, e-commerce, environment, healthcare, life science are covered. The list of topics spans all the areas of modern intelligent systems and computing.

The publications within "Advances in Intelligent Systems and Computing" are primarily textbooks and proceedings of important conferences, symposia and congresses. They cover significant recent developments in the field, both of a foundational and applicable character. An important characteristic feature of the series is the short publication time and world-wide distribution. This permits a rapid and broad dissemination of research results.

Sang-Yong Rhee · Jooyoung Park
Atsushi Inoue
Editors

Soft Computing in Machine Learning

Springer

Editors
Sang-Yong Rhee
Kyungnam University
Gyeongnam
Korea

Atsushi Inoue
Eastern Washington University
Washington
USA

Jooyoung Park
Korea University
Seojong
Korea

ISSN 2194-5357 ISSN 2194-5365 (electronic)
ISBN 978-3-319-05532-9 ISBN 978-3-319-05533-6 (eBook)
DOI 10.1007/978-3-319-05533-6
Springer Cham Heidelberg New York Dordrecht London

Library of Congress Control Number: 2014933545

Printed on acid-free paper

Springer is part of Springer Science+Business Media (www.springer.com)

Preface

As users or consumers are now demanding smarter devices, intelligent systems are revolutionizing by utilizing machine learning. Machine learning as part of intelligent systems is already one of the most critical components in everyday tools ranging from search engines and credit card fraud detection to stock market analysis. You can train machines to perform some things, so that they can automatically detect, diagnose, and solve a variety of problems. The intelligent systems have made rapid progress in developing the state of the art in machine learning based on smart and deep perception. Using machine learning, the intelligent systems make widely applications in automated speech recognition, natural language processing, medical diagnosis, bioinformatics, and robot locomotion. This book aims at introducing how to treat a substantial amount of data, to teach machines and to improve decision making models. And this book specializes in the developments of advanced intelligent systems through machine learning. It consists of 11 contributions that features illumination change detection, generator of electronic educational publications, intelligent call triage system, recognition of rocks at uranium deposits, graphics processing units, mathematical model of hit phenomena, selection and mutation in genetic algorithm, hands and arms motion estimation, application of wavelet network, Kanizsa triangle illusion, and support vector machine regression. Also, it describes how to apply the machine learning for the intelligent systems. This edition is published in original, peer reviewed contributions covering from initial design to final prototypes and verifications.

To help readers understand articles, we describe the short introduction of each article as follows;
1. "Sudden Illumination Change Detection and Image Contrast Enhancement": This paper introduces a new method for detecting sudden illumination change efficiently in real time by using local region information and fuzzy logic. Also it proposes an automatic parameter selection method for contrast limited adaptive histogram equalization method by using entropy of image and adaptive neural fuzzy inference system.

2. “Frame concept for generator of electronic educational publications”: This paper describes a system of methods to create the generator of Electronic Educational Publications (EEP) for higher education on credit education technology, in order to automate a process of EEP.
3. “Intelligent call triage system with algorithm combining decision-tree and SVM”: This article proposes a new online learning algorithm that constructs a subspace with a decision tree for a call triage support system. This algorithm can build a subspace with a decision tree for learning an increasing number of call triage records online.
4. “Recognition of Rocks at Uranium Deposits by Using a Few Methods of Machine Learning”: For uranium extraction in Kazakhstan, this paper proposes “adjacency cube” method for integration of results of few interpretation algorithms. Data interpretation can be performed using learned systems, such as artificial neural network (ANN), Linear Discriminant Analysis Classifier (LDAC), Support Vector Classification (SVM), k-Nearest-Neighbor (k-NN) and etc.
5. “More Faster Self-Organizing Maps by General Purpose on Graphics Processing Units”: As there are many loops which are simple calculation in learning algorithm of SOM, this paper describes a benefit of GPGPU to accelerate learning of Self-Organizing Maps(SOM). GPGPU feature an acceleration of the specific use computation and an adequtability for acceleration of simply calculation of loop iteration.
6. “Analysis of Local Concerts using Facebook Adapting the Mathematical Model of Hit Phenomena”: This paper presents a mathematical model for the hit phenomenon in entertainment within a society as a stochastic process of interactions of human dynamics. This model can be applicable to Japanese motion picture market or numbers of Facebook viewers.
7. “Effective Selection and Mutation in Genetic Algorithm for PAPR Reduction of OFDM Signal”: This paper proposes an improved genetic algorithm (GA) for the PAPR reduction based on tone injection (TI). The effectiveness of the GA method is demonstrated by numerical simulations in terms of PAPR, computation time and bit error rate (BER).
8. “Hands and arms motion estimation of a car driver with depth image sensor by using particle filter”: This paper proposes to incorporate some particles intentionally switching the left and the right of the hands in a framework of particle filter. This idea can reduce the mistaken (opposite) determination of left and right and will increase the opportunity to recover automatically from the opposite determination.
9. “SNR Improvement of an Optical Wave Microphone Using a Wavelet Network”: As the optical wave microphone can detect the sound without making any contact with others, the authors adopted wavelet network as information processing. This paper verifies a noise reduction at low frequency using only proposed algorithm.
10. “Origin of Kanizsa triangle illusion”: This paper introduces Kanizsa triangle figures equipped with stripe notches on its inducers to experimental participants.

The research concludes considering the mechanism of Kanizsa triangle illusion for global functional property.
11. "LTS-SVMR for Modeling of Nonlinear Systems with Noise and Outliers": This article proposes the least trimmed squares support vector machine regression (LTS-SVMR) which is the hybrid of the least trimmed squares (LTS). LTS-SVMR can effectively remove large noise and outliers under the proper initial nonlinear function and ensure robust performance.

We would appreciate it if readers could get useful information from the articles and contribute to creating innovative and novel concept or theory. Thank you.

Sang-Yong Rhee
Jooyoung Park
Atsushi Inoue

Contents

Sudden Illumination Change Detection and Image Contrast Enhancement

Bayanmunkh Odgerel and Chang Hoon Lee

Department of Electronic Engineering,
Paichai University Daejon, Korea
{pcu_b,naviro}@pcu.ac.kr

Abstract. Sudden illumination change is considered as one of crucial issue in many computer and robot vision (CRV) applications. Most CRV algorithms fail when sudden illumination change occurs; therefore, it is essential that images to be detected then the illumination change occurred images need to be enhanced in order to keep the appropriate algorithm processing in real time. We introduce a new method for detecting sudden illumination change efficiently in real time by using local region information and fuzzy logic. The effective way for detecting illumination changes in lighting area and the edge of the area is to analyze the mean and variance of the histogram of each area and in order to reflect the changing trends previous frame's mean and variance for each area of the histogram is used as an input. The changes of mean and variance make specific patterns when sudden illumination change occurs. Fuzzy rules were defined based on the patterns of the input for detecting illumination changes. Proposed method was tested with different datasets through the evaluation metrics; in particular, the specificity, recall and precision have showed high rates. We also proposed an automatic parameter selection method for contrast limited adaptive histogram equalization method by using entropy of image and adaptive neural fuzzy inference system. The results showed that the contrast of images could be improved. The proposed algorithm is robust to detect both global sudden illumination change and big moving object, and it is also computationally efficient in real time applications.

Keywords: Fuzzy Logic, Fuzzy Rule, Region Information, Image Histogram, Adaptive Neural Fuzzy Inference System, Contrast Limited Adaptive Histogram Equalization.

1 Introduction

Nowadays many computer and robot vision (CRV) applications including moving object detection and tracking, foreground extraction, surveillance system, and many other algorithms fail when sudden illumination change occurs. This is the key issue in image processing branch. Specifically there arise many noises caused by illumination change [1]. Illumination changes occur frequently in images due to many light sources such as shadows, switching light on or off, opening or closing a curtain,

S.-Y. Rhee et al. (eds.), *Soft Computing in Machine Learning*,
Advances in Intelligent Systems and Computing 273,
DOI: 10.1007/978-3-319-05533-6_1,

reflection of light from surface and the variation of sunlight during the day [2] in our everyday life. Sudden illumination change affects the entire or particular pixels in the same way and scene which changes its contrast globally (or partially) from one frame to another. Our research work concentrated on global illumination changes. When sudden illumination change occurs most computer and robot vision algorithms lose their information suddenly. B.Xie et al [3] utilized order consistency for sudden illumination change detection. This method only considers reference and current frame's neighboring pixels and used threshold. The sudden illumination changes lead to misunderstanding of background and foreground regions [4]. Our previous work [5] only used the changes of mean and variance of histogram of whole frame. That case our method has detected the whole sudden illumination change occurred frames. After illumination change occurred image has detected images need to be enhanced in order to retrieve the information that has lost through illumination change. Ulugbek et al [6] proposed the method that enhances the image contrast. But that method uses reference image to enhance the images.

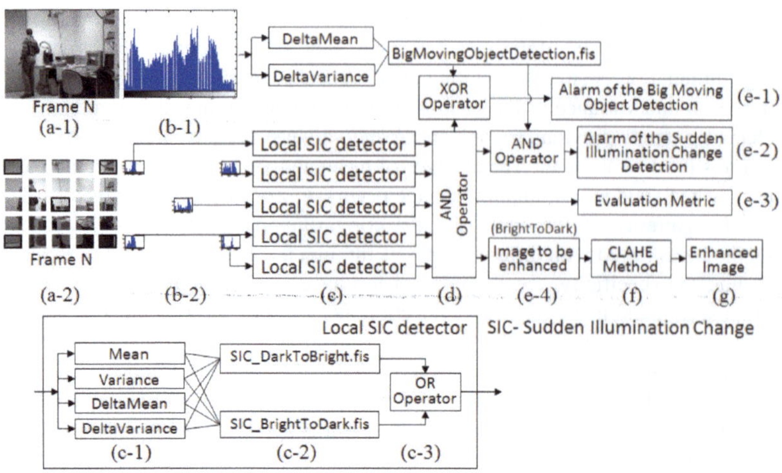

Fig. 1. Proposed sudden illumination change detection system

2 Preliminaries

In this section, we review the background of methods that were used in our work. Broad histograms reflect a scene with significant contrast (as shown in Fig.2.a) and value of that histogram's mean and variance is big, whereas narrow histograms reflect less contrast (as shown in Fig.2.b) and corresponding mean and variance is small.

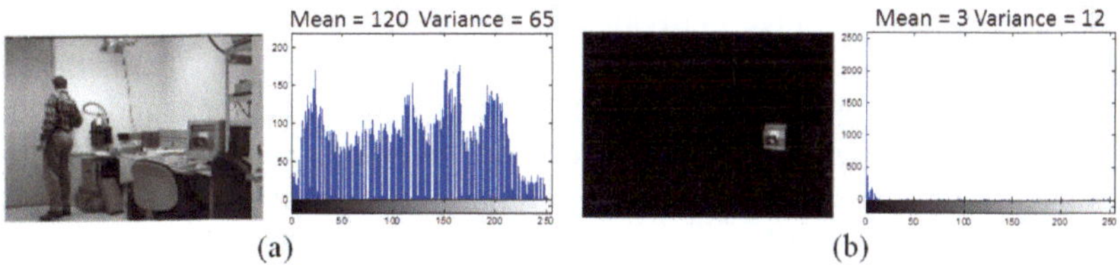

Fig. 2. Turning off light case of consecutive frames and its histogram

2.1 Fuzzy Logic

Fuzzy sets and logic were introduced by L.Zadeh in 1965. Fuzzy logic has been used to improve decision-making, reduce resource consumption, and increase performance [7]. Heart of the fuzzy logic is fuzzy inference system (FIS). In this research we designed fuzzy inference system for detection of sudden illumination change. Adaptive Neural Fuzzy Inference System (ANFIS-first introduced by Jyh Shing et al [8]) integrates the best features of fuzzy systems and neural network [9]. It has an advantage of learning from data.

Fuzzy rule base consists of a set of linguistic statements, called rules. These rules are of the form IF premise, THEN consequent where the premise is composed of fuzzy input variables connected by logical operators (e.g. AND, OR, NOT) and the consequent is a fuzzy output variable. Similarly to how membership functions are defined, the fuzzy rule is derived either based on domain knowledge, or through using machine learning techniques [7].

2.2 Contrast Enhancement Method

In image processing, enhancing the contrast of image is essential part. Histogram equalization is common technique by distributing original histogram of image equally for enhancing contrast of image. Drawback of this technique is it amplifies the noise and makes artificial (or strange) effects. Contrast limited adaptive histogram equalization (CLAHE) is a technique for avoiding the excess amplification, while maintaining the high dynamic range of the sub-block [10] by clipping histogram of image. Clipping parameter should be defined by user manually. Therefore it cannot be applied in real time by defining the parameter through user manually.

3 The Proposed Method

This paper has main two contributions. The first contribution is set of fuzzy rules for changes of the mean, variance, delta mean and delta variance (local region information) of histogram of edges and lighting areas. Work flow for proposed sudden illumination change and big moving object detection method depicted in Fig.1.

The second contribution is automatic parameter selection method for CLAHE method. By enhancing image contrast which is illuminated by light source we can keep track information that is in interest.

Sudden illumination change is divided in two cases one can occur from bright to dark and the other can occur from dark to bright. Input image Fig.1 (a) is divided into sub-regions and five regions are chosen; including tiles of edges and middle one. Height and width of frame is divided into equal five tiles, respectively. In Fig.1 (b) histograms of each selected tiles are depicted. By differentiating mean and variance of histogram of frame N-1 and N we can find the delta mean and delta variance, respectively. Mean, variance, delta mean and delta variance of histogram are used as crisp inputs into fuzzy inference system called SIC_DarkToBright.fis and SIC_BrightToDark.fis as shown in Fig.1 (c-2).

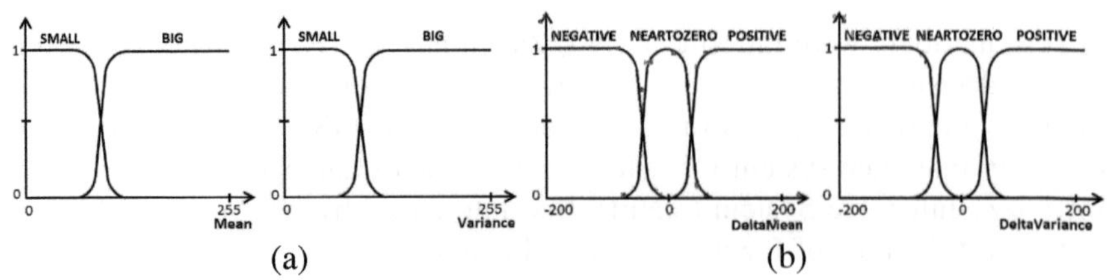

Fig. 3. Fuzzy membership function and fuzzy sets

Detection result of two FIS are combined with OR operator. Changes of mean, variance, and delta mean and variance make specific patterns when sudden illumination change occurs. Detection results of selected tiles are combined with logical AND operator (Fig.1.f) in order to check the illumination change on a whole region (frame). Once illumination change (only for bright to dark case) has detected CLAHE method was applied on the frame where sudden illumination change occurred from bright to dark case. Mean and variance could be the element of fuzzy set "Small" or "Big" as shown in Fig.3 (a). If histogram's mean and variance is small then those values belong to fuzzy sets "Small", otherwise those values belong to fuzzy set "Big", delta mean and delta variance could be the element of fuzzy set "Negative", "NearToZero" or "Positive" as shown in Fig.3 (b) , respectively.

3.1 Fuzzy Rule Expression for Change Detection

In general, histogram is broad (uniform or equal) while if image contrast is normal as shown in Fig.2 (a) and histogram is narrow (or not equal) while image contrast is too dark or too bright as shown in Fig.2 (b). Thus mean and variance of histogram follows the contrast of image directly. The mean and variance of histogram make specific patterns depending on the image contrast.

A change of local region information generates specific form when sudden illumination change occurs. Fuzzy rules were defined based on forms of that input. Simulation work for detection of sudden illumination change is depicted in Fig.4. First five rows show mean, variance (first column), delta mean, variance (second column) and detection result (third column) of local region. Sixth row indicates that the alarm for final detection of sudden illumination change.

Changes of delta mean and variance makes specific patters when moving object (especially big) covers the frame. Fuzzy rules were defined based on changing patterns of that input. In this work we detect big moving object using global region information inferring logical operators with local region information. In left most part of bottom of Fig.5 global region information (mean, variance, delta mean and variance) is illustrated and detection result of big moving object and consecutive frames are shown bottom of the figure. Specifically, from frame number 796 object starts covering the frames and our method detect the object that is moving.

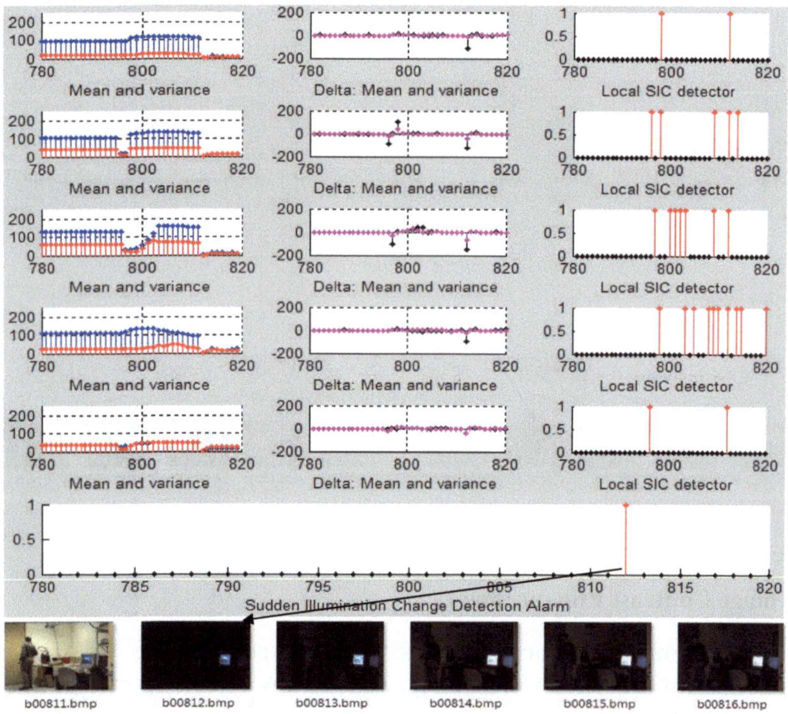

Fig. 4. Detection of sudden illumination change using local region information

3.2 Global versus Local Region Information

In our proposed method local region; including area of edges and middle one are selected as highlighted in black squares in Fig.1 (a-2). Calculating histogram of small region spends less time than whole region's histogram calculation and it enables efficiency of our proposed method. Selected tiles are enough to represent the global illumination change of whole frames. Reason is when global illumination change occurs on a certain frame then scene contrast changes its color globally. Another reason is to get rid of camouflage effects. Our previous work [5] have used the variations of mean and variance of global region's histogram was utilized for detecting sudden illumination change occurred frames.

Variation of mean and variance for big object covering and sudden illumination change were similar to each other as shown in Fig.5. With global region information it was hard to distinguish whether sudden illumination change or big moving object. Hence this causes the false positive of our detection system. Thus five tiles are chosen as shown in Fig.4 and Fig.6. By using local region information we can distinguish big moving object and sudden illumination change. Detection result of sudden illumination change using local region information and big moving object using global region information are inferred using logical operators in order to distinguish them. By using local region information it is unable to detect big moving object as shown in Fig.7.

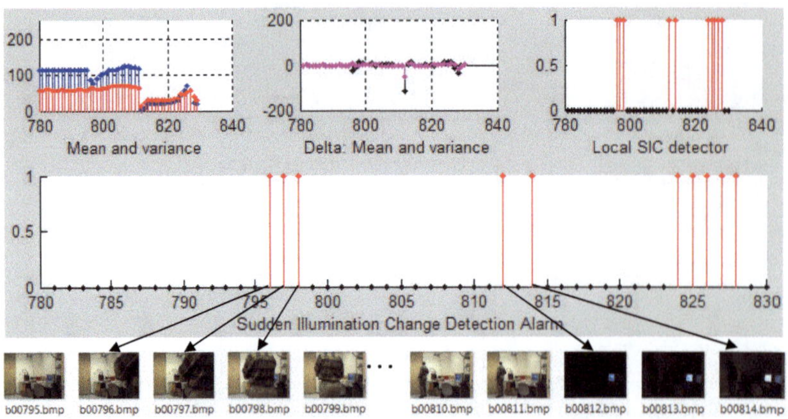

Fig. 5. Experiment result of the global region information

3.3 Image Contrast Enhancement

Once sudden illumination change has detected on a certain frame, CLAHE method was applied to enhance the contrast of image which is caused by change of sudden illumination. Work [6] has proposed image contrast enhancement method based on local histogram specification. In this method reference image is required in order to enhance the contrast. We proposed the automatic clip limit parameter selection for CLAHE by utilizing entropy of image and ANFIS. Our contrast enhancement method does not require any reference image and work adaptively depending on entropy of image which is trained into ANFIS. Zujun Hou et al [11] investigated the connection between entropy and image visibility. The entropy was used to measure the visibility of image in our work. It keep changes depending on image contrast as shown in Fig.8. Entropy of images is plotted as a graph on bottom of the Fig.7. Entropy of image and several manually selected parameter (for example entropy = 7.1571, clip limit = 0.0034 etc.) are used by doing some experiment in order to train the data into ANFIS. Work flow for above method is depicted in Fig.8.

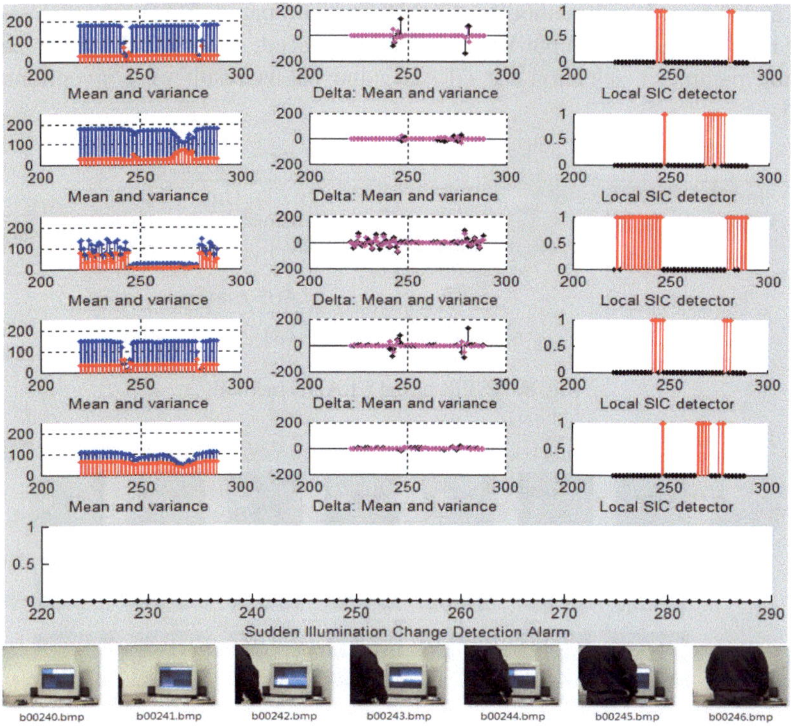

Fig. 6. Experiment result of the local region information

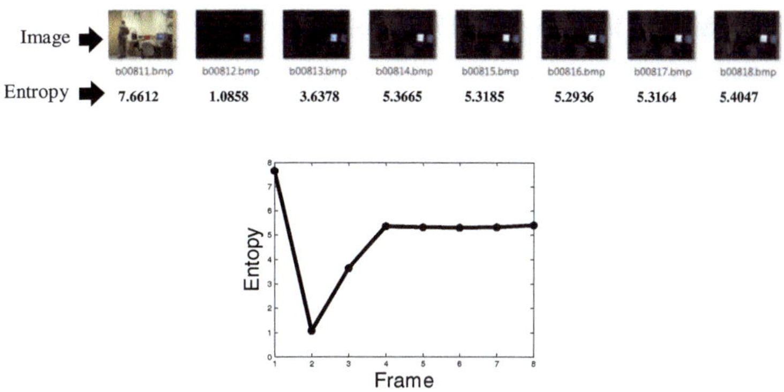

Fig. 7. Image and its entropy

In Fig.8 (a) input image (on which sudden illumination change occurred from bright to dark case) and entropy of that image is calculated (b), several stipulated input-output data is used in to ANFIS and CLAHE assigns its clip limit parameter from ANFIS (d). In Fig.9 result of conventional adaptive histogram equalization

(AHE) and our proposed method for CLAHE are compared. Fig.9 (a) indicates entropies of input images, (b) input images to be enhanced, (c) is result of our proposed automatic parameter selection for CLAHE and (d) is result of conventional AHE method.

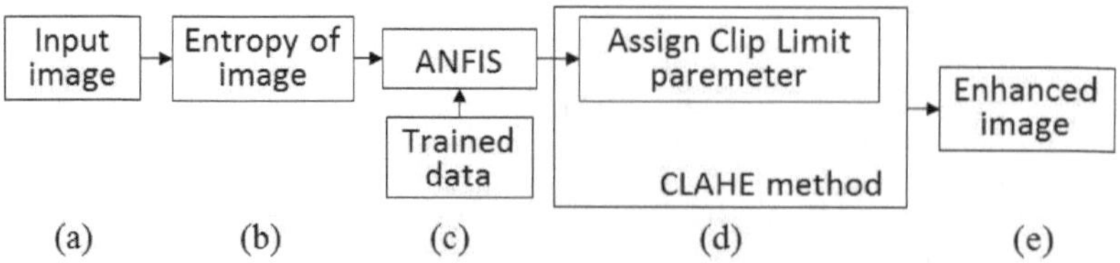

Fig. 8. ANFIS based CLAHE method

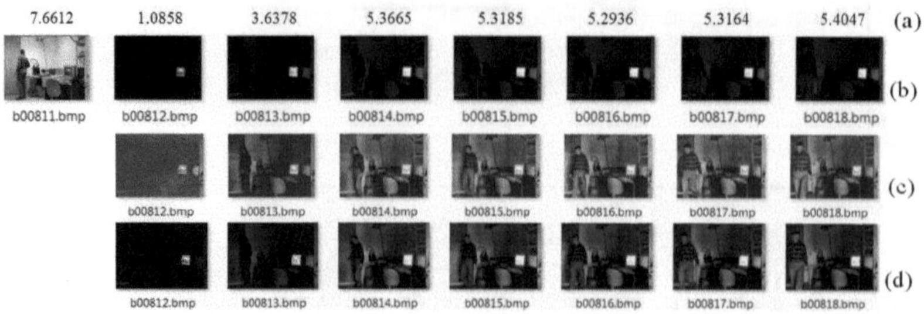

Fig. 9. Comparison result of conventional AHE and our proposed method

4 Experiment Result and Discussion

In this study, some experimental results dealing with the proposed method was provided by using delta mean and variance in order to detect sudden illumination change and big moving object on certain frames. Our method performed with different kind of benchmark datasets (*.avi video sequences) such as LightSwitch, Camouflage and simulation work was done in Matlab.

The result of sudden illumination change was tested with different evaluation metrics as defined in [12]. Specifically precision, recall and specificity showed high rates as shown in Table 2. In order to analyze the performance of our proposed method we tested with different evaluation metrics. Performance is good if specificity, precision and recall are high or equal to one. Evaluation metrics are compared with ground truth of illumination changes and detection result of our proposed method. In Fig.10 simulation of detection result of sudden illumination change and enhanced image are shown. On the detected frame CLAHE method was applied. Result of enhanced image shown in below of Fig.10.

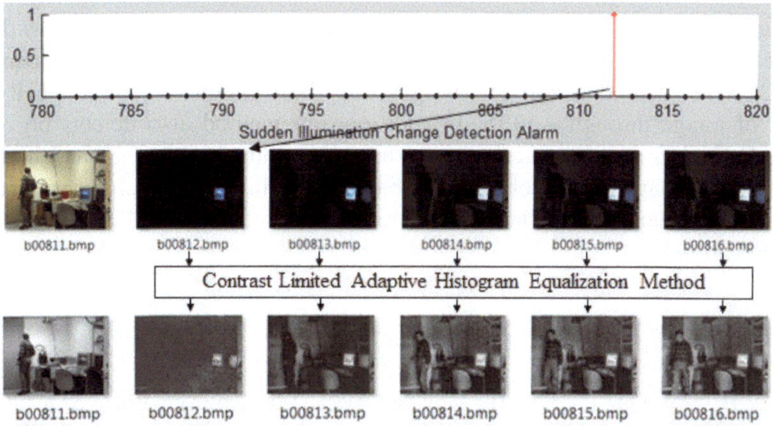

Fig. 10. Sudden illumination change detection and enhanced images

Table 1. Detection result with different evaluation metrics

Evaluation Metrics/ Dataset	LightSwitch1	LightSwitch2	Camera Parameter	Motinas_toni Change_ill
Total frame	271	27	499	339
Ground Truth of SIC	3	15	3	3
True positive	3	4	3	2
False positive	0	0	0	0
False negative	0	11	0	1
True negative	271	27	499	336
Recall	1	0.2	1	0.6
Specificity	1	1.0	1	1.0
FPR	0	0	0	0
FNR	0	0.0	0	0.0
PWC	0	0.3	0	0.2
Precision	1	1.0	1	1.0
FMeasure	1	0.4	1	0.8

5 Conclusion

In our previous work [5] only variations of mean and variance of global region's histogram was utilized for sudden illumination change detection. In this paper a new method for detection of sudden illumination change was introduced by using local region information and fuzzy logic approaches. Fuzzy logic improves the efficiency to detecting the sudden illumination change and it enables introduced method can processes online. CLAHE method was applied after illumination change has detected in order to enhance the contrast of image in prior to the application of other methods. Thus, CLAHE method adaptively assigns clip limit parameter automatically using entropy of image through ANFIS. In addition, our method also detects big moving objects. Detection result has tested with different evaluation metrics and high rates. Our method is adaptive and robust to global sudden illumination changes. The result shows that the identified method is computationally efficient and applicable in real time applications.

References

1. Le, V.H., Mai, Q.H., Lee, C.-H.: Moving Object Detection from Video Sequence Using Optical Flow Estimation and Wavelet Decomposition. In: Proceedings of KIIS Spring Conference 2012, vol. 22(1) (2012)
2. Bascle, B., Bernier, O., Lemaire, V.: Learning invariants to illumination changes typical of indoor environments: Application to image color correction. International Journal of Imaging Systems and Technology 17(3) (October 2007)
3. Xie, B., Ramesh, V., Boult, T.: Sudden Illumination Change Detection Using Order Consistency. Image and Vision Computing 22(2), 117–125 (2004)
4. Choi, Y.-J., Lee, J.-S., Cho, W.-D.: A Robust Hand Recognition In Varying Illumination. In: Advances in Human-Computer Interaction, ch. 4, pp. 53–70. InTech Education and Publishing (October 2008)
5. Bayanmunkh, O., Lee, C.-H.: Sudden Illumination Change Detection Using Local Region Information and Fuzzy Logic. In: Proceedings of KIIS Spring Conference 2013, vol. 23(1) (2013)
6. Khusanov, U., Lee, C.-H.: Image Enhancement Based on Local Histogram Specification. Journal of Korean Institute of Intelligent Systems 23(1), 18–23 (2013)
7. Kapitanova, K., Son, S.H., Kang, K.-D.: Using fuzzy logic for robust event detection in wireless sensor network
8. Jyh-Shing, Jang, R.: Adaptive Network-Based Fuzzy Inference System. IEEE Transactions on Systems, Man, and Cybernetics 23(3) (May/June 1993)
9. Patnaik, S., Yang, Y.-M.: Soft Computing Techniques in Vision Science
10. Ritika, R., Kaur, S.: Contrast Enhancement Techniques for Images A Visual Analysis. International Journal of Computer Applications 64(17), 0975–8887 (2013)
11. Hou, Z., Yau, W.-Y.: Visible Entropy: A Measure for Image Visibility. In: 2010 International Conference on Pattern Recognition (2010)
12. Evaluation metrics, `http://www.changedetection.net`

Frame Concept for Generator of Electronic Educational Publications

Altynbek Sharipbay, Assel Omarbekova, and Alma Zakirova

Dept. of Theoretic Computer Science,
L.N. Gumilyov Eurasian National University
2 Mirzoyana Street, Astana, Republic of Kazakhstan
{sharalt,omarbekova,alma_zakirova}@mail.ru

Abstract. This work describes a system of methods to create the generator of Electronic Educational Publications (EEP) for higher education on credit education technology. The purpose is to develop technology which would make it possible to automate stages of EEP creation.

An Electronic Educational Publications (EEP) is a set of digital, text, graphic, audio, video and other types of information which have means of programmed control and documentation, possess adaptive properties and intelligent interface and can be thus placed in the global Internet network.

This work was supported by a grant under the program "Targeted development of university research focused on innovative results" for the project "Methodology, algorithms and programs for electronic educational publications generation".

Keywords: e-learning systems, higher education, intelligent tutoring systems, intelligent electronic textbook, generator, knowledge base.

1 Introduction

Currently many high-quality software products designed for learning process purposes are presented at the market of computer-based learning systems. These products are developed both by domestic and foreign companies (mainly foreign). As for development of EEP, it is quite a laborious task. While creating any EEP, in addition to selection of educational material, the developer has to consider some common issues, like data representation, implementation of user interface, preparation of control tests, etc. The EEP Generator allows us to automate these actions.

The Generator is designed for educators' independent work. It allows the instructors to create EEPs for their subjects in accordance with the requirements of national standard of the Republic of Kazakhstan ST RK 34.017-2005 "Information technology. Electronic publications. Electronic educational publications." There are no counterparts of this shell program in Kazakhstan.

S.-Y. Rhee et al. (eds.), *Soft Computing in Machine Learning*,
Advances in Intelligent Systems and Computing 273,
DOI: 10.1007/978-3-319-05533-6_2, © Springer International Publishing Switzerland 2014

2 Background

2.1 Related Works

Using the Generator reduces the time of creation of electronic educational publications and expands the range of potential authors.

There is no need in programming when creating e-textbooks with the Generator. The teacher puts all the training materials in the program, and then it generates the EEP on the basis of the loaded data. Input and output data of the Generator is shown in Figure 1.

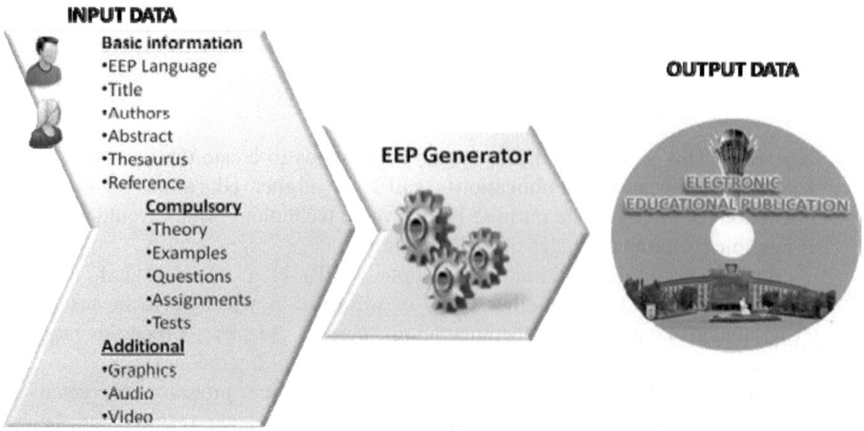

Fig. 1. Input and output data of the Generator

The advantages of this approach are as follows:

- Economically efficient;
- Allows concentrating on the material selection;
- Reduces the time of creation.

Data input is maximally simplified. There are two ways to input text information (theory, examples, assignments, questions and tests).

The first is to enter data in the EEP by selecting the buttons in the generator "Theory", "Examples", "Assignments," "Questions", "Graphics", "Audio," "Video", "Tests". The required structure of folders and files is created automatically and saved in a database.

The second method. Let's say you have already generated some data. To connect it to the EEP, you should just bring the information to a specific format.

The foundation of the generator development is the frame concept, designed by Marvin Minsky, one of the founders of the artificial intelligence theory. He believed that mental processes are based on people's memories stored in different data structures – frames [1]. Thanks to them, one realizes visual images (visual frames), understands the words (semantic frames), arguments and actions (scenarios frames and behavioral models). Thus, frame is an elementary semantic unit. This concept is used to improve the perception and control attention of students. For this purpose, EEP provides the possibility of installing educational material as a sequence of so-called "screens" – audio-visual frames, or simply frames. Training material is divided into semantic frames. Typically, one screen contains one semantic frame. If the semantic frames are simple to understand, hierarchically equal in the training material and follow each other, they can be placed on one frame-screen. This method of information presentation can significantly improve the perception of the information. Failure to follow these rules dramatically worsens the perception of information. For example, if one frame-screen contains several certain concepts, torn by explanations, the information cannot be perceived (violation of the equivalence of hierarchical frames). The correct choice would be to place concepts on the first frame-screen and their explanations on the next frames (preferably, each explanation should be on its own frame).

2.2 The Hierarchical Structure of the Generated EEP

The collection of frames, which simulates some subject area, is a hierarchical structure in which the frames are gathered by genus-species relations. The hierarchical structure of the generated EEP is shown in Figure 2.

The EEP consists of a cover page, abstract, table of contents, elements of learning and management functions. An element of learning is a semantic unit of training information. Elements of learning are lessons, modules and components. The lesson is the least semantic unit of training information, which includes the following parts: content, questions, assignments, tests, a glossary and a reference.

The content is the theoretical part of the learning material presented in the lesson. The content must consist of text, graphics, audio, video and other information related to the theme and appropriate to the selected audience, goals and objectives of the EEP.

Fig. 2. The hierarchical structure of the generated EEP

Management functions are registration, navigation, viewing, testing, learning, designing and help content. Registration provides the ability to enter data about the user (student) in order to keep statistics about him/her. Navigation is the apparent relationship between the elements of the EEP and provides the necessary means of orientation and movement in the EEP. View provides a view of the total volume of teaching material included in the EEP. The user will not perform the assessments. Testing allows you to check the student's knowledge as of the current lesson, and throughout the course of study. In order to do that, the tester randomly selects questions from a common database of questions on curriculum, provides questions and answers related to the chosen topic. Test results will be displayed on the screen [2].

Education provides the student with the opportunity to explore the theoretical material on the current lesson, look at the questions and answers, to do exercises or solve problems and take the tests. In case of insufficient number of correct answers to the tests, the student cannot go to the next lesson and will continue to examine the current lesson.

Start date begins with the selection of learning paths, which can be defined in three ways: manual selection, test selection and a full selection.

The manual selection allows the tutor or the student to determine the trajectory of their own by selecting blocks, modules, and lessons in the index of the EEP.

In the test selection the program automatically defines the trajectory of learning based on the test results throughout the volume of educational material included in the EEP. In this case, the trajectory will include only the lessons, which were failed by the students.

The full selection includes the entire learning trajectory of educational material (all classes, modules and components) included in the EEP.

One may continue the learning process according to the selected trajectory only, taking lesson by lesson.

The designer is a toolbar that facilitates the creating the path of learning.

2.3 The Node in the Framing System

Each node in the framing system has the form shown in Figure 3.

Concept (Frame Name)		
Slot 1	Slot value 1	Procedures
Slot 2	Slot value 2	Procedures
...	...	...
Slot N	Slot value N	Procedures

Fig. 3. The node in the framing system

The concepts in every node are defined as a set of attributes and their values that are contained in the frame slot. In addition to the specific value the slot may store procedures and rules that are invoked if it is necessary to calculate this value. Among them there are procedures-daemons and procedures-slaves. The first start automatically when a certain condition arises, while the latter are activated only by a special request.

A slot is an attribute associated with the node in a system based on frames.

The slot is a component of the frame. Slot is named in accordance with the type of attribute; the value of the slot may be an instance of the attribute, the other frame or a facet [3,4].

Each slot can be linked to one or more procedures that are performed when values of slots change.

Most often the following procedures are linked to the slots:

1. If added (performed when new information is added to the slot);
2. If deleted (this is done, when some information is deleted from the slot);
3. If needed (this is done, when requested information from an empty slot).

The data structure of a frame lesson is presented in Figure 4.

Procedure 1: daemon type procedure "if needed", the necessary file is retrieved from the database, the program runs the macros that converts the documents into an *.htm format to be included to the EEP. When delivering the text information, the generator adjusts the text information of all classes to a single style: font Times New Roman (Cyrillic), size 14, the background of the main text is white, correct answers' text background is light blue, hints' background is light yellow.

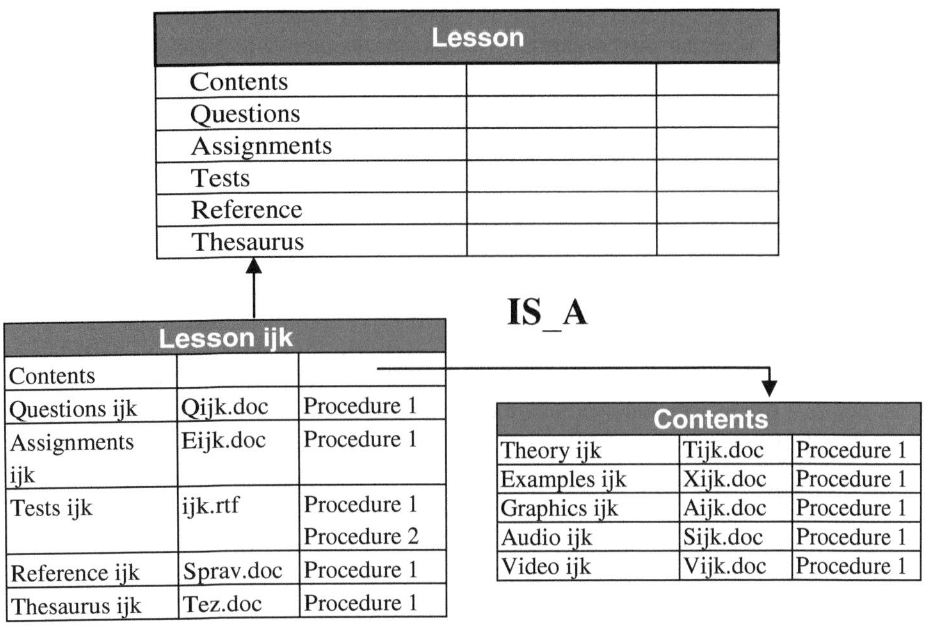

Fig. 4. The data structure of a frame lesson

Procedure 2: if the test is not failed the student is moved to the next lesson. Once all the lessons of this unit are successfully completed, the student is automatically taken to the next unit. To move to the next block, you have to first pass the intermediate control of knowledge (the number of tests equals to the number of lessons in the unit; each test is randomly chosen for every lesson). To move to the next module, you have to pass the endpoint control of knowledge. At the end of training the final control of knowledge is offered. In case of an incorrect answer to a specific question in the intermediate, endpoint or final testing this lesson is included into the learning program again [5].

By the types of perception of information people are divided into three types: visual (perception of information through sight prevails), auditory (through hearing) and kinesthetic (through experience). The perception of information of all types of people is improved if all channels are used. Didactic feature of the EEP that are created by the generator is its multisensory experience. The student receives all kinds of information at the same time. In addition, the transition from one screen (and, hence, the semantic unit) to another is controlled by the student (just press the button). This allows the student to choose the pace that is comfortable for him/her. Besides, it adds the kinesthetic element in the perception of information.

To improve the perception the program also uses a number of methods of attention management. These include the high amount of graphic illustrations, animations, fonts, and color selection, etc.

Using the concept of frames, multisensory experience, and attention management possible technologies enabled the creation of EEP having high density of information. The learning process becomes an individual process (for teaching in a computer room or at a personal computer at home).

Psychologists have noticed that the recognition of objects is easier in the usual context than in a precarious situation. That is, the student will be easier and faster perceiving new information from various disciplines using the EEP with a unified interface, a single way of composition of the learning elements and management structure, all of which can be easily created with the generator [6].

3 Implementation

There exist a number of general tool systems for creation of electronic textbooks, such as: NVU, KompoZer, TurboSite, Microsoft Learning Content Development System (LCDS), iSpring Pro 6, the ADONIS, UROK, AOSMIKRO, STSENARIY, APOLLO, META, Multimediya-Brig, Costoc, LinkWay, Quest, TenCore, etc.

In contrast to the existing systems, the developed EUI generator has the following distinctive features:

- EUI formation according to the state standard ST of RK 34.017-2005 "Information technology. Electronic publication. Electronic educational publication";
- hierarchical structure of the EUI: modules, blocks, lessons;
- the following elements of training are distinguished: theory, examples, exercises, questions and answers, tests, graphics, audio, video, thesaurus, directory;
- multilingual support;
- existence of a system of knowledge control;

- multimedia opportunities;
- personified trajectory of training;
- logging of trainees' actions;
- administration;
- highly simplified data input.

After program installation in the Start-up menu 4 labels will appear: "Registration", "Formation", "Textbook", "Help". After registration of the software product the developer will be able to create an EUI.

3.1 Authoring System

Upon selection of "Formation" from the menu the authoring system intended for independent development of EUI by the teacher will be started. The teacher needs to enter into the generator input data, such as the EUI language, heading, abstract, thesaurus, directory and also mandatory data including theory, examples, questions, tasks, tests, etc. (Figure 5).

Fig. 5. EEI formation

3.2 User System

After formation of EEI is complete, it is possible to see the result by selecting "Textbook" tab in the main menu. The title of the developed EEI where the output data and elements of management of EEI are located will open.

"Authors", "Summary", "Help" tabs reflect those data which were entered when forming the contents. Pressing the "Table of contents" button opens the window visually reflecting the textbook's structure.

Pressing the "Contents" button allows trainees to choose an operating mode (Figure 6).

The first, display mode. This mode allows viewing only training materials. Thus, access to tasks, questions and graphics will not be allowed.

The second, testing mode. In this mode the training program provides testing upon all training materials. After completing the test it is possible to see test results.

The third mode, starting the training. To start training, the trainee needs to be registered.

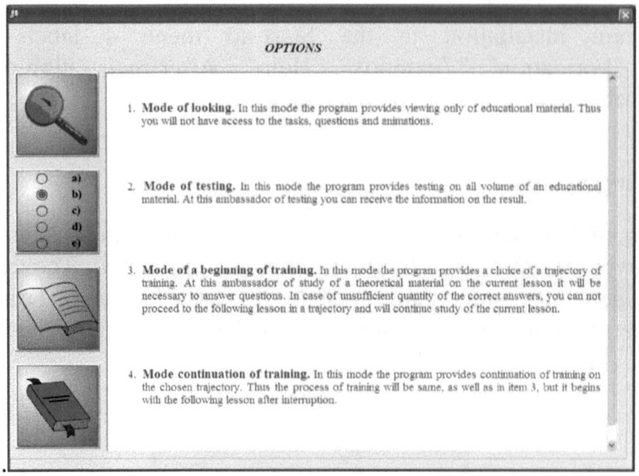

Fig. 6. Operating mode selection

In this mode the training program provides a choice of a trajectory of training (figure 7). Thus, after studying of a theoretical material it will be necessary to answer the questions for the current lesson. In case the number of the correct answers is insufficient, the trainee won't be able to go on to the next lesson in the trajectory of training and will continue studying the current lesson. Beside the current testing, other forms of testing are provided, namely intermediate testing (upon transition to the next block), transitional («rubezhny») (upon transition to the next module) and final (upon the completion of training).

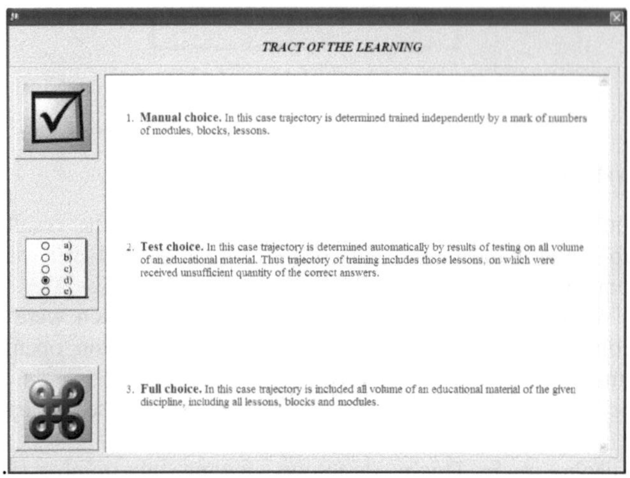

Fig. 7. Formation of a trajectory of training

The fourth mode is continuing training. In this mode the training program provides continued training within the chosen trajectory. The process of training resumes from the lesson following the one completed previously.

The mode of the beginning of training allows to choose one of the three trajectories of training: manual selection, selection by testing and full selection.

Manual selection implies that the trajectory is defined by the trainee by selecting the respective numbers of modules, blocks and lessons.

Selection by testing allows for automatic selection of a trajectory based upon the results of testing on all training material. In this case the trajectory of training includes only those lessons where the number of correct answers is insufficient.

Full selection guides the trainee through all training materials in the discipline, including all lessons, modules and blocks.

After trajectory is defined the user passes directly to the training session (fig. 8) .

This window allows access to theory, examples, tasks, questions, multimedia, the directory and the thesaurus of the first lesson in the training trajectory. For transition to the next lesson it is necessary to correctly answer test questions of the current lesson.

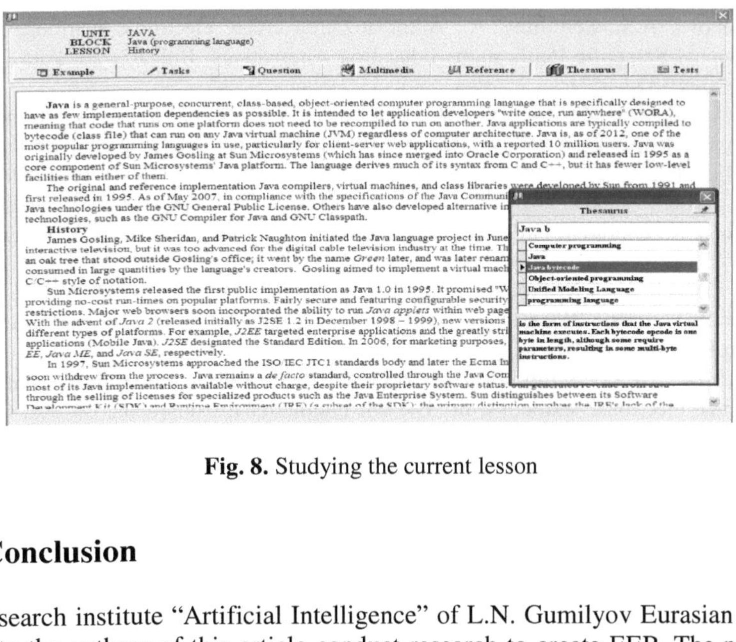

Fig. 8. Studying the current lesson

4 Conclusion

In the research institute “Artificial Intelligence” of L.N. Gumilyov Eurasian National University the authors of this article conduct research to create EEP. The process of creating EEP has achieved progress as it has been through the following steps so far:

- there has been developed a national standard for the creation and introduction of EEP 'to the educational process;
- there has been developed technology of automated creation of local EEP;
- there has been developed technology of automated creation of online EEP;
- with the help of the EEP generator the teaching staff have elaborated a number of EEP in different disciplines, which are used in the learning process, as

evidenced by acts of application. Some EEP are posted on the official website of the Information Technology Faculty of L.N.Gumilov ENU www.fit.enu.kz (e-content project). There are certificates of state registration of intellectual property obtained for the EEP generators and for the developed EEP [7].

Currently we have implemented a Kazakh speech synthesizer to voice content of EEP. We are also working to provide the EEP voice-activated management function. For the control of knowledge we are planning to connect up a subsystem of the expert system based on fuzzy logic.

Detailed information on automated technology of EEP creation is available at www.e-zerde.kz/generator.

References

1. Minsky, M., Frames, M.: Energy (2009)
2. Abdymanapov, S.A., Sharipbaev, A.A., Omarov, A.N., Baymuratova, G.G., Nurgujin, M.R., Baygelov, K.G., Aljanov, A.K., Omarbekova, A.S.: ST RK 34.017-2005, Information technology. Electronic publications. Electronic educational publications
3. Omarbekova, A.S., Seifullina, A.: Automatization create electronic learning. Journal of International Scientific Publications Education Alternatives, Bulgaria 10, Part 1, 242–250 (2012) ISSN 1313-2571
4. Omarbekova, A.S., Sharipbay, A.A., Zakirova, A.B., Barlybayev, A.: Generation of intelligent e-learning editions on the basis of ontologic model. Journal of International Scientific Publications Education Alternatives, Bulgaria 11, Part 1, 4–11 (2013) ISSN 1313-2571
5. Lapshin, V.A.: Ontologies in computer systems, p. 222. Scientific World, Moscow (2010)
6. Gruber, T.R.: The role of common ontology in achieving sharable, reusable knowledge bases / / Principles of Knowledge Representation and Reasoning. In: Allen, J.A., Fikes, R., Sandewell, E. (eds.) Proceedings of the Second International Conference, pp. 601–602. Morgan Kaufmann (1991)
7. Omarbekova, A.S., Seifullina, A.: Building automation ontological generation of intellectual electronic textbooks. In: The 6th International Conference on Soft Computing and Intelligent Systems, Kobe, Japan, pp. 2217–2219 (2012)

Intelligent Call Triage System with Algorithm Combining Decision-Tree and SVM

Kento Yokose and Tomoki Hamagami

Graduate School of Engineering, Yokohama National University
79-5 Tokiwadai, Hodogayaku, Yokohama, Japan
k.yokose.ynu@gmail.com

Abstract. In this paper, we propose a new online learning algorithm that constructs a subspace with a decision tree for a call triage support system. A call triage is an operation to determine the level of first-aid service quality on the basis of the severity and urgency of a victim. The call triage support system used in Yokohama is operated smoothly. However, further improvement in the accuracy detection is an issue because of the high percentage of misclassifications. This issue has been discussed in our prior implementation of a Bayesian network call triage support system. However, results indicate that the accuracy is decreased by over-training on the unlearned data. In this paper, we propose an algorithm to build a subspace with a decision tree for learning an increasing number of call triage records online. The evaluation experiment uses the past call triage records. Its results show that the proposed method can judge a call triage efficiently.

Keywords: Call Triage, Decision tree, Support vector machine.

1 Introduction

Recently, against the background of a super aging society, an increasing number of ambulance calls have led to delays in the ambulance response. The average ambulance response time in 1998 was 6 min [1]; this increased to 8.2 min [2] in 2012. This sharp increase in the ambulance response time adversely affects the quality of emergency medical care.

In order to overcome the issue, a call triage system has attracted attention in the field of public healthcare. A call triage is a classifying decision approach that prioritizes patients' treatments on the basis of the severity of their condition, for determining emergency service efficiently.

The operation has had far-reaching effects on the reduction of the cost of emergency service without decreasing the quality of such service.

However, according to the bias in a fail-safe environment, the incorrect decisions called under-triage classifications pose other problems, which adversely affect both the optimal use of limited emergency resources and the performance of the triage.

In our prior research, we utilized a Bayesian network algorithm to enhance a call triage support system [3]. The results revealed an increase in the classification

S.-Y. Rhee et al. (eds.), *Soft Computing in Machine Learning*,
Advances in Intelligent Systems and Computing 273,
DOI: 10.1007/978-3-319-05533-6_3, © Springer International Publishing Switzerland 2014

accuracy. However, we also confirmed that over-training decreased the detection performance of the remaining training data during the structural learning of the Bayesian network.

Although the amount of call triage data is increasing by the day, the new triage data was beyond the scope of our prior research, and it is difficult to improve the call triage system without using these new data.

In order to improve the performance of the call triage system, the following should be investigated:

- Improvement of the accuracy of a call triage with unknown data, and
- Online learning for utilizing new call triage data.

This paper proposes a new call triage support system that uses a support vector machine (SVM) and a decision tree. The proposed method reduces the ratio of over-triage, improves the accuracy, prevents the over-training, and supports the newly added call triage data adaptively.

2 Call Triage Support System

A call triage is a type of decision-making support system that determines the priority of patients' treatments, the order and priority of emergency transport, and the transport destination for the patients according to the severity of their conditions. For instance, the typical call triage data used in this paper consist of 87 items such as gender, age, respiratory problems, neurological problems, and the Gold Standard (GS) of severity score obtained by professional medical diagnoses. The observed data consist of 86 items that exclude those of GS from the call triage data.

Table 1 shows the GS values that correspond to the call triage categories (A+, A, B, C+, and C) used in Yokohama city. In the current call triage decision, the underestimation of the severity of a patient's condition is called under triage (UT). UT causes delayed treatment and may lead to the loss of the chance to save a patient's life. In particular, the fatal under triage (FUT) decision for patients who should be tagged to A+ or A to C, must be avoided. Furthermore, failure cases with respect to the classification of patients who should be tagged to B to C are called serious UT (SUT). Meanwhile, the overestimation is usually called over triage (OT). OT increases the unnecessary use of emergency resources and the corresponding cost, and decreases the quality of rescue service.

Table 1. Category Classification Used in Yokohama City

Category	GS	Definition
A+	Death or sober (cardiac arrest)	Life crisis is highly imminent.
A	Sober (non-cardiac arrest)	Life crisis is imminent.
B	Serious	There is potential threat to life.
C+	Mild (non-self-walking)	There is low potential threat to life, but it is difficult to transport the patient.
C	Mild (can walk on their own)	There is low potential threat to life, and it is easy to transport the patient.
I	Nothing	The call triage can be disabled.

The essential role of a call triage is to avoid FUT and reduce OT. The running call triage support system used in Yokohama city uses a combined algorithm that consists of a multivariate learning logistic model and a medical-expert-based “Pickup method” [4]. In order to avoid FUT, the combined algorithm tends to bias a fail-safe decision. The bias poses a risk of increasing OTs.

Our prior research applied a Bayesian network for a decision algorithm to solve the above issue. Although it enabled an improvement in the accuracy of classification, over-training occurred during the structural learning of the Bayesian network and reduced the detection performance to the remaining training data. Moreover, the classification of the newly increased call triage data was beyond the scope of the combined algorithm and our prior proposal. In order to realize the trade-off, a new call triage algorithm is required to utilize the newly increased triage data adaptively for classification enhancement.

In response to these issues, in this paper, we propose a mechanism to construct a subspace at each call triage decision and make a call triage decision. This is reflected in the subspace of the new call triage data. Further, a call triage decision is taken in the subspace. In addition, the machine learning method that can build a model with high identification capability even for the unknown call triage data is required for taking the call triage decision in the subspace.

To construct this mechanism, in this paper, we propose new call triage support system that uses a support vector machine (SVM) and a decision tree.

3 Call Triage Support System Using Support Vector Machine and Decision Tree

The proposed method consists of the following two processes:

Constructing "subspace" by decision trees

The call triage data that are neighbors of the observation data that are the target for making a call triage decision are used for creating a subspace to measure the proximity of qualitative data such as call triage data, using a decision tree.

The decision tree is an algorithm that uses a tree structure for classification corresponding to the input data. The categorical value of a call triage is allocated to the leaf (end) node. The other nodes are allocated with the conditional decision regarding the observation data.

If the observed data that are the target for making a call triage decision are given, build a subspace for the call triage data that have arrived in the same leaf node as the observed data.

Learning of "super-plane" by SVM in the subspace

A call triage decision super-plane of the SVM is built for the observed data in a subspace constructed by a decision tree. The model represents high discrimination capability even for a small number of samples in the subspace created by using the SVM.

3.1 Construction Subspace by Decision Tree

The call triage data used in this paper are $\{\mathbf{x}_i, y_i\}$ $(i = 1, \ldots, n)$. $\mathbf{x}_i$ denotes the observed data. y_i denotes GS. n represents the amount of call triage data.

Upon the construction of a subspace, a decision tree is used as the means of measuring the proximity of qualitative data such as the call triage data. We developed a decision tree using the CART algorithm and the teacher's value y_i.

$\mathbf{x}^d$ denotes the observation data that are the target for making a call triage decision. Se implies that $\mathbf{x}^{sd}$is classified as $\mathbf{x}_i$ in the same leaf node as that of $\mathbf{x}^d$ in all the constructed decision trees. Then, a subspace $\mathbf{T} = \{\mathbf{x}_1^{sd}, \ldots, \mathbf{x}_j^{sd}, \ldots, \mathbf{x}_m^{sd}\}$ was built using $\mathbf{x}^{sd}$. m denotes the amount of observation data belonging to subspace $\mathbf{T}$.

3.2 Parameter of Subspace

A subspace was built in the decision tree, learning was performed again in another subspace by the learner, and a triage decision was made. However, whether the reliability of any of the determination result outputs could be improved could not be confirmed. Therefore, it is necessary to quantify the evaluation of the determination result output in order to provide a mechanism to determine the user's judgment.

Assume that DV is the judgment value of the case in which the decision tree makes the call triage decision for $\mathbf{x}_j^{sd}$. By using DV, we define the evaluation value of the judgment result.

Evaluation of subspace

The call triage data belonging to the subspace determine the category value corresponding to the leaf node DV. l denotes the amount of call triage data $\mathrm{DV} = y_j^{sd}$. The evaluation of subspace $\mathbf{T}$ is as follows: $\mathrm{U} = \frac{l}{m} * 100$.

Evaluation of FUT

When reaching a call triage decision with respect to the call triage data belonging to $\mathbf{T}$ by using a decision tree, we observe that an FUT occurs. In this case, $\mathbf{T}$ has any anomalous data that cause FUT. Therefore, $\mathrm{E} = \{true, false\}$ is an indicator of the presence of anomalous data in the considered dataset. If $\mathrm{E} = true$, it is shown that the anomalous data that induce FUT include $\mathbf{T}$.

3.3 Classifier in Subspace

Then, we need to execute the learning algorithm from the data samples limited in the subspace. However, as mentioned before, neither the logistic model nor the Bayesian network approach enables the user to obtain a high accuracy with a small amount of training data. In order to overcome the issue, in this study, we used the SVM, which could train using few samples, provide a high learning accuracy, and avoid the overtraining problem.

SVM is known as a kind of pattern classifier that can form a hyperplane between the call triage categories on the basis of a given sample. In the SVM, there is an objective function to maximize the margin; it calculates the distance between the pluralities support vector and the hyperplane.

In the subspace, SVM learns and reaches the triage decision in the following case:

- Category value of the call triage data contained in the subspace is two and over (i.e., $\mathrm{U} < 100$).

On the other hand, SVM does not learn in the following cases:

- Category value of the call triage data contained in the subspace is only one (i.e., $\mathrm{U} = 100$).
- There are no observation data $\mathbf{x}_i$ that arrived by following the same leaf node $\mathbf{x}^d$ in all decision trees.

At this time, the call triage decision for $\mathbf{x}^d$is reached in the decision tree.

4 Call Triage Decision Experiment

As mentioned before, this study constructs the subspace and derives the likelihood of the subspace for a call triage decision. There are 61,530 call triage data elements recorded in Yokohama and used in our experiment. GS is used as the judgment value. The parameters of decision trees are determined empirically and are shown in Fig. 1.

- Depth of decision tree for categories except category C+: 3
- Depth of decision tree for category C+ and the other categories: 10

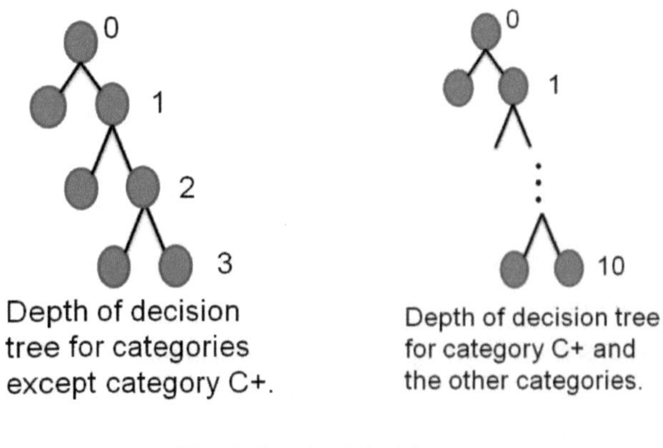

Fig. 1. Depth of decision tree

The parameters of the SVM are as follows:

- Kernel function: RBF kernel
- Gamma value: 0.0068

Based on the actual operation conditions, the evaluation condition of the percentage accuracy, percentage OT, and percentage UT is defined as follows:

1. Constructing a subspace **T** .
2. Learning decision hyperplane in spatial learning target **T** with SVM.
3. Making the call triage decision for new data $\mathbf{x}^d$.

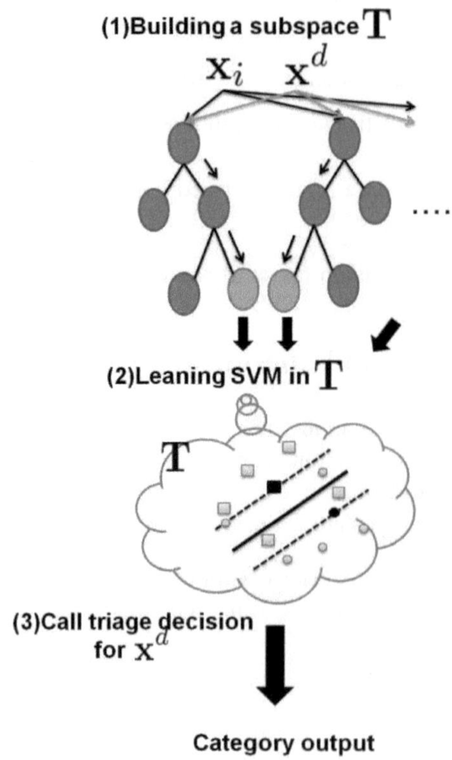

Fig. 2. Flow of call triage discriminant

The call triage dataset is divided into 10 sets; 1 set is for the test, and the remaining 9 sets are for the training.

4.1 Call Triage Decision Only If Building a Subspace

In this section, we will discuss the call triage decision in the subspace without the evaluation values mentioned in Section 3.2.

The results of 10 cross-validation tests are shown in Tables 2 and 3.

Table 2. Discriminant results

Discriminant results (number)						
		A+	A	B	C+	C
GS	A+	1608	15	22	690	3
	A	24	12	25	732	14
	B	60	16	92	3266	88
	C+	269	104	359	47843	0
	C	0	0	1	0	6287

Table 3. Number of severe or dead patients in category C

Algorithm	*Number of FUT*
Proposed method	17
Bayesian Network	5
Logistic	1

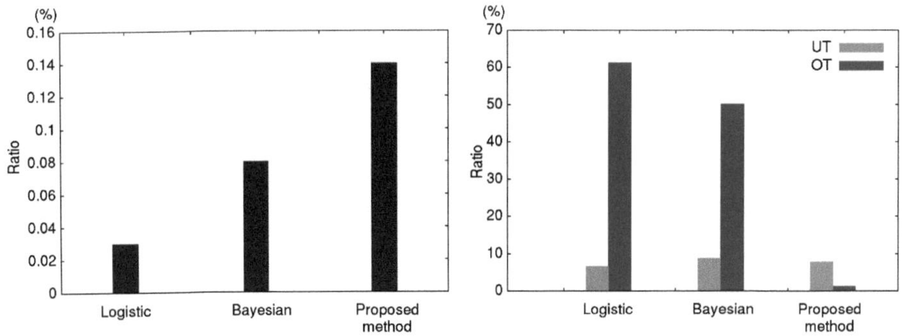

Fig. 3. Accuracy of call triage (left: SUT, right: UT and OT)

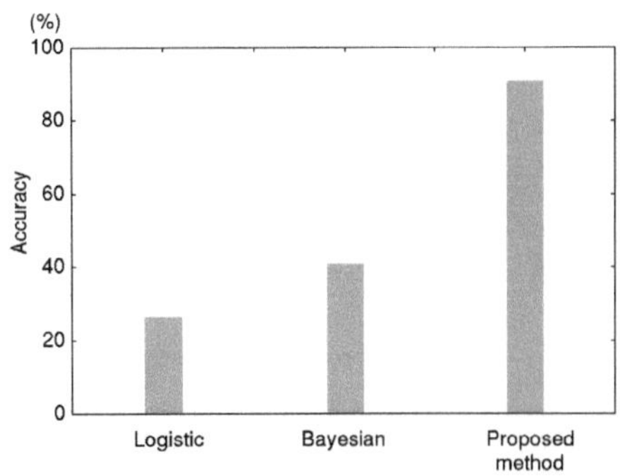

Fig. 4. Accuracy of call triage

In Table 2, the vertical category denotes GS, and the horizontal category denotes the output from the proposed method. There are 1608 call triage records of category A+, which are determined to be A+ on the basis of GS. Table 3 shows the number of FUT cases observed in the case of each method. Figure 3 shows the percentages of SUT, UT, and OT. Figure 4 shows the ratio of the matched determination values to the GS for all instances.

Compared with the logistic-model-combined algorithm and our Bayesian-network-based prior proposal, the currently proposed method could significantly improve the percentage of OT (for instance, an increase by 1.35% over that of the prior proposal). Consequently, the overall accuracy has improved to 90%. On the other hand, occurred 17 to FUT. However, 17 cases of FUT were observed in the result of the proposed method. In the proposed method, the number of FUTs increased with the priority of

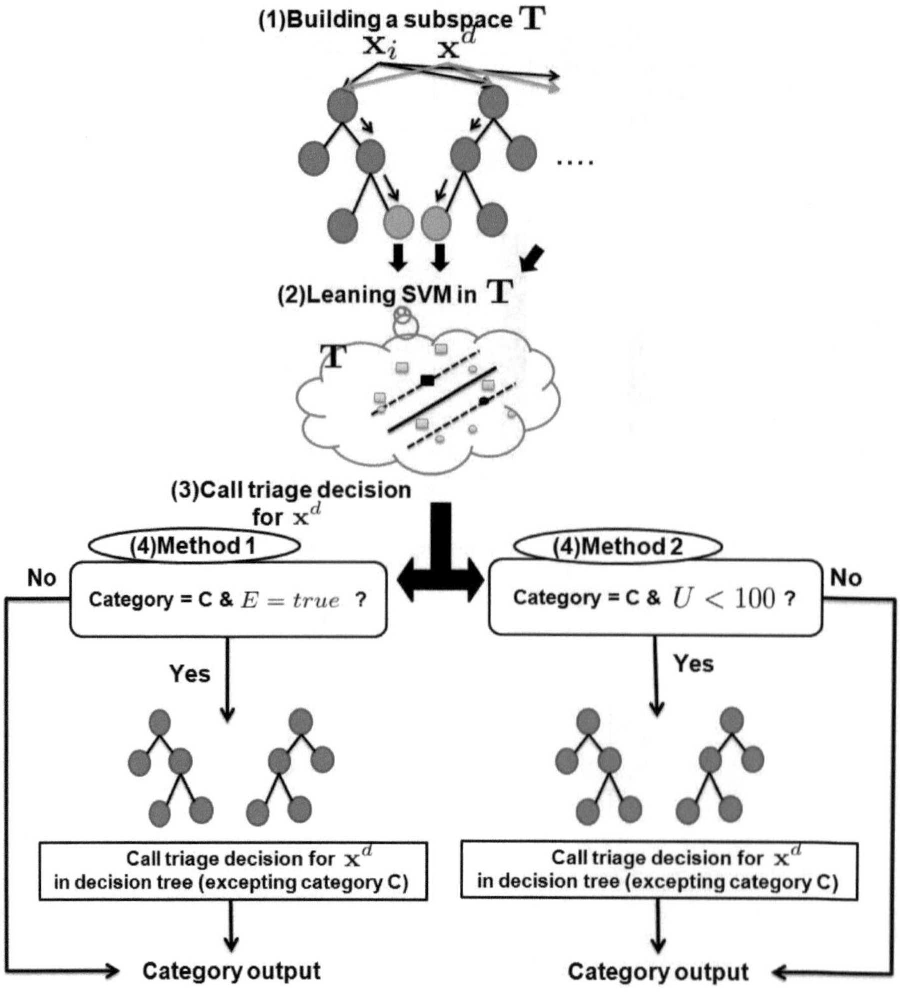

Fig. 5. Modified flow diagram of call triage discriminant

accuracy. Therefore, mechanisms to satisfy the actual condition are required. Thus, if the decision value is the output category C, we aim to satisfy the actual condition by modifying the decision value of the spatial accuracy U, E . The reject decision value of category C depends on U, E , ensuring a fail-safe call triage decision.

4.2 Call Triage Decision Using Evaluation Values

In order to reduce FUT in the proposed method, an experiment was conducted considering the spatial accuracy. The modified algorithm is as follows:

1. Constructing a subspace $\mathbf{T}$.
2. Learning decision hyperplane in spatial learning target $\mathbf{T}$ with SVM.
3. Reaching the call triage decision for new data $\mathbf{x}^d$.
4. Modifying a decision value by using U and E as follows:

Method 1: Does not output category C of $\mathbf{T}$ such as $\mathrm{E} = true$, and call the triage decision for $\mathbf{x}^d$ in the decision tree (limited A+, A, and B).

Method 2: Does not output category C of $\mathbf{T}$ such as $\mathrm{U} < 100$, and call the triage decision for $\mathbf{x}^d$ in the decision tree (limited A+, A, and B).

In order to evaluate the modified algorithm, the experiments were conducted using Methods 1 and 2. Tables 4, 5, and 6, and Figs. 6 and 7 show the results of 10 cross-validation tests.

Table 4. Discriminant results (Method 1)

	Discriminant results (number)					
		A+	A	B	C+	C
GS	A+	1,608	15	25	690	0
	A	24	12	35	732	4
	B	60	16	167	3,266	13
	C+	269	104	359	47,843	0
	C	2	0	5,121	0	1,165

Table 5. Discriminant results (Method 2)

	Discriminant results (number)					
		A+	A	B	C+	C
GS	A+	1,608	15	22	690	0
	A	24	12	38	732	1
	B	60	17	176	3,266	3
	C+	269	104	359	47,843	0
	C	2	0	6,189	0	97

Table 6. Number of severe or dead patients in category C

Algorithm	*Number of FUT*
Proposed method (Method 1)	4
Proposed method (Method 2)	1
Bayesian Network	5
Logistic	1

Compared with the logistic-model-combined algorithm and the Bayesian network, both the proposed methods 1 and 2 enable us to significantly reduce the percentage of OT (9.09 and 11.42%, respectively.) Furthermore, the overall accuracy improved to 80%.

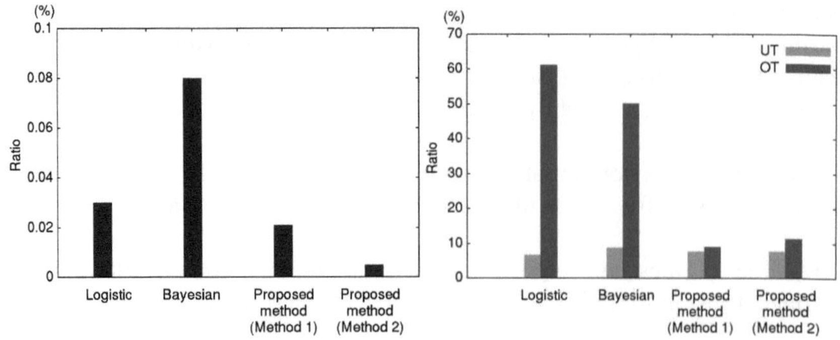

Fig. 6. Precision result of call triage (left: SUT; right: UT and OT)

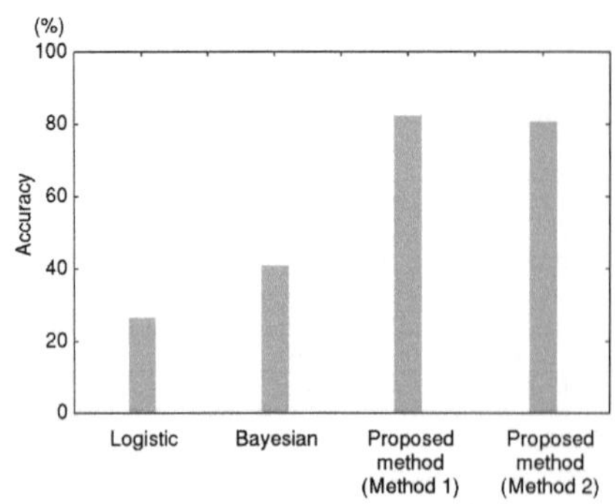

Fig. 7. Accuracy of call triage

The ratio of UTs was almost the same in each method. The ratios of the SUT in the case of the proposed methods were 0.021% and 0.0049%, respectively. Further, the ratios of the FUT in the case of the proposed methods were improved slightly compared with the other conventional methods. The number of FUT cases was 4 (Method 1) and 2 (Method 2).

Compared with that of Method 2, the accuracy of Method 1 was better. Meanwhile, the sufficiency extent of such as the numbers of FUT and the ratio of SUT is better Method 2. It was not only output category C in the subspace containing any anomalous data in Method 1. On the other hand, it was not output category C in the subspace such as $U < 100$. If any anomalous data were included in the subspace, naturally $U < 100$. Further, it was $U < 100$ even when there were multiple types of category values of the call triage data present in the subspace. As a result, Method 2 had fewer cases of the output category C in the subspace than Method 1. Therefore, Method 2 satisfied the actual condition better than Method 1.

These results indicated that the proposed methods contributed to an improvement in the call triage accuracy and the realization of fail-safe services.

5 Conclusion

This paper proposes a high-accuracy call triage system by an algorithm combining a decision tree and an SVM. By constructing a subspace using the decision trees, the proposed algorithm reduces the computational cost and improves accuracy of the call triage system. Furthermore, by using the SVM to avoid the over-training problem, the algorithm can construct a hyperplane for dividing the call triage categories.

The evaluation experiment shows that the proposed decision algorithm enables the user to improve the decision accuracy without increasing the number of fatalities under triage. Moreover, the proposed algorithm enables us to utilize a new dataset of emergency services for constructing and growing the system. If the amount of call triage data increases, it is possible to include the increased amount of call triage data in the subspace by constructing the subspace for every call triage decision.

As future work, we plan to build an intellectual emergency medical support system (SMART-ER) for application in the real world.

References

1. Fire and Disaster Management Agency, Fire fighting White Paper 1999 edition, GYOUSEI (2000)
2. Fire and Disaster Management Agency, Fire fighting White Paper 2012 edition, Katsumi Print (2013)
3. Yunoki, S., Hamagami, T., Oshige, K., Kawakami, C., Suzuki, N.: High-accuracy Decision of Call-triage by using Bayesian Network. The Transactions of the Institute of Electrical Engineers of Japan. C, A Publication of Electronics., Information and Systems Society 132(1), 61–67 (2012)
4. Oshige, K., Kawakami, C., Mizusima, S., Moriaki, Y., Suzuki, N.: Evaluation of an algorithm for estimating a patient's life threat risk from an ambulance call. BMC Emergency Medicine (2009)

Recognition of Rocks at Uranium Deposits by Using a Few Methods of Machine Learning

E. Amirgaliev, Z. Isabaev, S. Iskakov, Y. Kuchin, R. Muhamediyev, E. Muhamedyeva, and K. Yakunin

Institute of Problems of Information and Control,
Ministry of Education and Science of Republic of Kazakhstan,
International Informational Technology University, Almaty, Kazakhstan

Abstract. Uranium extraction in Kazakhstan is carried out using underground leaching method. Economic performance depends on the production process speed and accuracy of geophysical data interpretation. Data interpretation can be performed using learned systems, such as artificial neural network (ANN), Linear Discriminant Analysis Classifier (LDAC), Support Vector Classification (SVM), k-Nearest-Neighbor (k-NN) and etc. In the paper "adjacency cube" method for integration of results of few interpretation algorithms is proposed. Learning algorithm for the "adjacency cube" with low computational complexity was developed. The proposed method improves quality of recognition by 2-3 percent.

Keywords: Geophysical research of boreholes, machine learning, artificial neural network, k-NN, uranium deposit, post-processing data, learning sample, "adjacency cube" method.

1 Introduction

This problem of classification of lithological types is weakly formalized by its nature, so its solution can be sought in the scientific field called machine learning. There is a wide variety of tasks and successful applications of machine learning.

The methods of machine learning are a broad class of algorithms: from decision trees, genetic algorithms, Linear Discriminant Analysis Classifier (LDAC), Support Vector Classification (SVM), k-Nearest-Neighbor (k-NN) up to artificial neural networks (ANN).

In the paper, a comparative analysis of different recognition algorithms applied to the logging data of uranium deposits in Kazakhstan was performed. The task of creating recognition system based on using multiple algorithms (which we refer to as "adjacency cube" method) was set, and also a simple algorithm for learning and recognition for the method at the stage of post-processing was developed. This approach increases quality of recognition by 2-3 percent.

The work consists of the following parts:

In the first part application of ANN to analyzing the logging data is considered.

S.-Y. Rhee et al. (eds.), *Soft Computing in Machine Learning*,
Advances in Intelligent Systems and Computing 273,
DOI: 10.1007/978-3-319-05533-6_4, © Springer International Publishing Switzerland 2014

In the second part the results obtained using different algorithms are compared.

In the third part an easy to compute method that can be applied at the stage of post-processing of data is proposed and the results are presented.

In conclusion the results are summarized and the aims of further research are formulated.

2 Applying ANN to Interpreting Geological Logging Data

Uranium extraction in Kazakhstan is performed by using the method of underground leaching, which is a less expensive and more environmentally-friendly extraction method. [1] Economic performance of the extraction process depends on the speed and accuracy of interpretation of geophysical data for which electrical methods are often used: induction logging (IL), apparent resistively logging (AR) and the method of the natural polarization (NP). The wide use of core testing in the analysis of rock is impossible, as a result of slowness of the process of obtaining the data.

Incorrect or inaccurate analysis of geophysical data leads to the loss of wells, excessive labor costs and reduces the economic indicators of production. Since the logging data interpretation is largely empirical by nature, there are no precise patterns; the possibility of applying learning systems, in particular, neural networks is evident. However, despite the advantages of ANN, the main of which is the ability of ANN to solve weakly formalized problems, there are significant problems met when using them, such as:

- The ambiguity in expert opinion
- The need for a large and equal number of examples from different classes
- Inability of the neural network to explain the result
- The necessity for careful preliminary preparation of data (removal of anomaly values, normalization, smoothing)

The above mentioned problems became evident in the first stage of setting up a neural network system for the interpretation of logging data on stratified-infiltration deposits of Kazakhstan [2-5].

During the research, neural network configuration was designed and the effect of the following factors was analyzed

- neural network learning algorithm;
- architecture of the neural network;
- method of normalization;
- smoothing methods;
- special methods of forming a training set;
- correcting logging data shift relative to each other

In total, there were about two thousand computational experiments. During the experiments, "a floating window of data" with the parameters 5+1+5 (five points above and below the current one, plus the current point) was applied to the input of the network of the three types of logging.

Note that the original data set is difficult to recognize and classify, as a result of that different classes of objects are mixed substantially in the feature space, and the number of different classes of objects differs tenfold (Fig. 1).

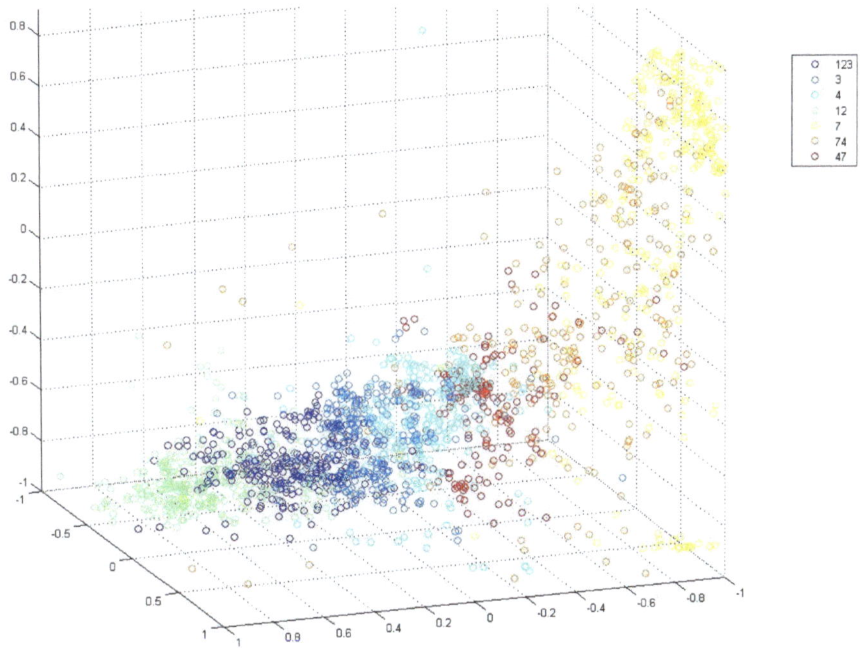

Fig. 1. Expert estimations in three dimensional (IL, AR, NP) feature space

Overall, the results of the preliminary research have shown that the use of ANN exclusively on array of data available to the researchers (about 12,000 samples) can achieve an average of 73% degree of successfully interpreted data when compared to the core trial (data obtained experimentally). There exists a spread of quality indicators in the interval of 5% (the worst result of recognition was 68.3%, the best one was 76.7%). Improving quality of interpretation requires the use of additional training samples and classification methods.

3 Comparative Analysis of Recognition Algorithms

Nowadays metric, logic (including fuzzy), Bayesian, statistic etc. classification methods have been developed and are being applied. Their use can generally improve the quality of the training set and improve recognition of data. In [6] the use of metric and statistical algorithms (Linear Discriminant Analysis Classifier (LDAC), Support Vector Classification (SVC) (Linear SVC and Non-linear SVC), Diagonal Linear Discriminant Analysis (DLDA), k-Nearest-Neighbor (k-NN)) is described and shown that they provide somewhat smaller average recognition accuracy compared with the artificial neural networks.

However, it can be assumed that synthesizing algorithms listed can in some way improve recognition. For further analysis, three algorithms were chosen: ANN, k-NN and recognition algorithm based on the analysis of the graph of the AR, which shows lower average result, however, defines impermeable rock well, which recognition is critical in the technology of uranium extraction. Comparison of results of the algorithms is shown in Table 1.

Table 1. Comparison of average results of algorithms

Borehole number	AR	ANN	k-NN
10172	0,615741	0,683333	0,650926
10174	0,613426	0,713735	0,546296
10175	0,646487	0,748132	0,720478
10176	0,741353	0,748132	0,720478
10177	0,533541	0,766771	0,636505
10197	0,686813	0,749608	0,77865
10198	0,593298	0,747144	0,617669
10217	0,621541	0,718025	0,679132
10229	0,678407	0,7366	0,729709
Average	0,6367342	0,734608763	0,675538348

4 "Adjacency Cube" Algorithm

The complexity of the above optimization problem makes it difficult to brute force the solution. However, we can offer some heuristics that will significantly reduce the computational complexity. Primarily, it is possible to combine the results of post-processing step.

Each of the classification algorithms (ANN, k-NN, AR) gives an average of 60% to 65% overlap with the responses obtained experimentally (by the core). Analysis showed that the results of the algorithms overlap in the following ratios:

AR - k-NN: 72.23%
KAR- ANN: 76.20%
k-NN - ANN: 77.78%

Moreover, each of the algorithms has different success rate and fault recognition rate for every class:

Table 2. Comparison of difference of algorithms

Algorithm	Recognition	1	3	4	7	59
AR	Correct recognition	0,919	0,663	0,043	0	0,743
	Fault recognition	0,216	0,379	0,878	0	0,435
k-NN	Correct recognition	0,788	0,689	0,214	0,055	0,2
	Fault recognition	0,182	0,439	0,734	0,822	0,767
ANN	Correct recognition	0,857	0,724	0,306	0,243	0,486
	ault recognition	0,122	0,366	0,714	0,662	0,541

It can be observed that it is possible to improve the overall result by combining the results of few algorithms (at the stage of post-processing) by choosing the most statistically appropriate response. A method referred to as "adjacency cube method" (ACM). ACM works on the basis of statistical data of recognition in two stages:

1. Training by creating the adjacency matrix.
2. Analysis of the results of recognition and selection of the final lithological type by using the adjacency matrix.

4.1 "Training" of the "Adjacency Cube"

Training of the ACM is held on the basis of data obtained in the course of the N algorithms interpretation. In this case, N is equal to three.

The first step is to fill the N-dimensional adjacency cube. The number of sides (dimensions) is equal to the number of algorithms and each side of the cube consists of a set of lithological types (Table 3).

Table 3. Codes of lithological types

Lithological code	Name of lithological type
1	Gravel, pebbles
3	Medium grained Sands
4	From small grained sands to super fine-grained sands
6	Aleurolite
7	Clay
47	Shortgrainedclayey sands
57	Clayey sandstone
59	The carbonatecement sandstone

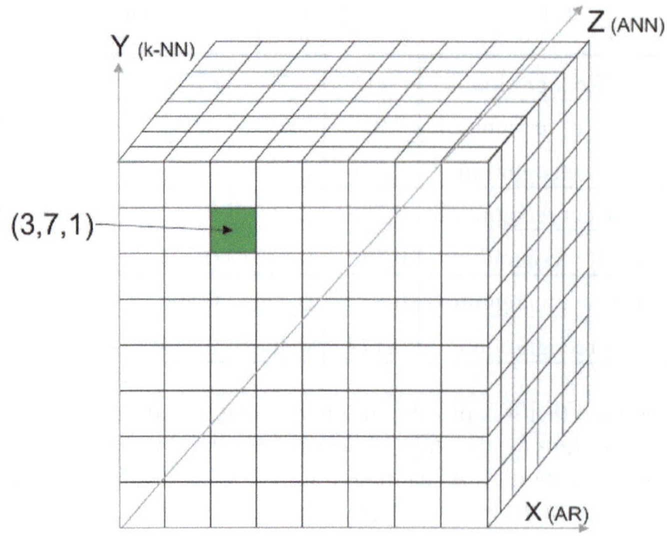

Fig. 2. Adjacency cube

Figure 2 shows a cube in graphical form where axis X is AR-algorithm, axis Y is algorithm k-NN and axis Z is ANN algorithm. Consider a cell with coordinates (3,7,1), where 3 corresponds to the third lithological type on AR-axis (X), 7-th along the k-NN-axis (Y), and 1-st along the ANN-axis (Z).

It is possible that some algorithms may not recognize certain classes. For example, the AR-algorithm does not recognize lithological type 47.

Filling the adjacency cube is performed as follows:

1. Choose every cell from the cube one-by-one
2. Go through the entire training set, and everywhere where it meets the set of results 3,7,1 (AR algorithms result is 3, k-NN is 7, ANN is 1) count the number of expert estimations for each lithological type (3,7,1).
3. Among the three possible lithological types, the one with maximum expert estimations is chosen and filled in the cell of the cube.
4. Go to the next blank cell of the cube.

In such a manner, we fill all the cells in a cube.

4.2 Analysis of the Results of Recognition and Final Selection of Lithological Type

Selection of the result among three proposed lithological type is performed using cube cells, filled during the first stage.

As the results of the experiments, up to 2-3% increase in quality of recognition was achieved (table 4).

Table 4. Results of the "adjacency cube" method

Borehole number	AR	ANN	k-NN	ACM
10172	0,615741	0,683333	0,650926	0,715741
10174	0,613426	0,713735	0,546296	0,731481
10175	0,646487	0,748132	0,720478	0,764574
10176	0,741353	0,748132	0,720478	0,764574
10177	0,533541	0,766771	0,636505	0,832293
10197	0,686813	0,749608	0,77865	0,797488
10198	0,593298	0,747144	0,617669	0,753998
10217	0,621541	0,718025	0,679132	0,769634
10229	0,678407	0,7366	0,729709	0,771822
Mean value	0,636734	0,73460876	0,67553834	0,766845
Mean square deviation	0,06	0,025	0,07	0,034

ACM achieves a better result than any of the algorithms separately for each borehole. At the same time the average result is higher than the best average (ANN) by 3.2% and the standard deviation is closest to the minimum among the three algorithms. Thus, the use of this algorithm improves the quality of recognition with a relatively small dispersion.

The computational complexity of the algorithm of forming the adjacency cube is linearly dependent on the amount of input data (O (n)), and the decision algorithm is a constant (O (1)).

5 Conclusion

Economic performance of the process of uranium extraction depends on the speed and accuracy of geophysical data interpretation. To interpret the data, as a result of poorly-formalized nature of the task, trained systems can be used, in particular, the ANN. A preliminary study showed that the use of ANN only achieves an average of 73% degree of successfully interpreted data compared to expert estimations. A comparison of metric and statistical classification algorithms with ANN indicates that these algorithms provide on average slightly worse results.

However, the quality of recognition algorithms for different rocks is different. This fact was the main in the proposed idea of "Adjacency cube method" (ACM) in the post processing stage. Application of ACM at the stage of post-processing can improve the result by 2-3 percent and decrease the variance of the results.

The time complexity of the algorithm of forming "Adjacency cube" method depends linearly on the amount of input data (O (n)), and the selection algorithm result is a constant (O (1)).

The aim of further research is the search and analysis of similar heuristics on the stage of pre-processing and synthesis methods over a wide range of algorithms that are designed to improve the quality characteristics of the system of recognition of lithological types in uranium deposits.

References

1. Jashin, S.A.: Underground acidic in situ leaching of uranium at the Kazakhstan deposits. "Gorny Zurnal" Scientific-Technical and Industrial Journal (3) (2008) ISSN 0017-2278
2. Muhamediyev, R.I., Kuchin, Y., Muhamedyeva, E.: Recognition of Geological Rocks At the Bedded-infiltration Uranium Fields by Using Neural Networks. In: 2012 IEEE Conference on Open Systems, Kuala Lumpur, 6 p. IEEE Xplore Database (2012) ISBN: 978-1-4673-1044-4, doi:10.1109/ICOS.2012.6417622, `http://cs.ieeemalaysia.org/ICOS2012/` (accepted August 2012)
3. Kuchin, Y., Muhamedyev, R., Muhamedyeva, L.: Interpretation of log data of boreholes. In: The 9th International Conference: Information Technologies and Management 2011, April 14-15. Information Systems Management Institute, Riga (2011) ISSN 1691-2489, 98 c
4. Muhamediyev, R.I., Kuchin, Y., Muhamedyeva, E.: Analysis of the data of geophysical research by means neuronet methods. In: The Proceedings of the VIII International Scientific Conference on Electronics And Computer Technologies, IKECCO 2011, pp. 289–297. SDU, Almaty (2011) ISBN 9965-792-10-0
5. Muhamediyev, R.I., Kuchin, Y., Muhamedyeva, E.: Geophysical Research of Boreholes: Artificial Neural Networks Data Analysis. In: The 6th International Conference on Soft Computing and Intelligent Systems and the 13th International Symposium on Advanced Intelligent Systems, SCIS-ISIS 2012, Kobe, Japan, November 20-24, pp. 825–829 (2012) 978-1-4673-2743-5/12/$31.00 ©2012 IEEE
6. Amirgalieyev, Y., Kuchin, Y., Iskakov, S., Muhamedyev, R., Muhamedyeva, E.: Evaluation of the quality of the neural network recognition of lithologic layers on uranium deposits. In: Proceedings of the Scientific-Practical Conference, Actual Problems of Informatics and Control, pp. 262–270. Institute of Problem Informatics and Control, Almaty ISBN 978-601-06-2094-

More Faster Self-Organizing Maps by General Purpose on Graphics Processing Units

Shinji Kawakami and Keiji Kamei

Nishinippon Institute of Technology, 1-11 Aratsu Kanda,
Miyakogun, Fukuoka, 800-0394, Japan
{s130103,kamei}@nishitech.ac.jp
http://www.nishitech.ac.jp/~kamei

Abstract. Recently, GPUs have been attempted to process for general computation, it is called "General Purpose Graphic on Processing Units(GPGPU)" and is the focus of attention. The feature of GPUs is that small and limited capable processors are parallelized on a massive scale, and those processors operate synchronously. From this feature, GPGPU have capability of acceleration of the specific use computation. The capability of GPGPU is specifically suitable for acceleration of simply calculation of loop iteration because processing is in parallel.

In this paper, we adopt GPGPU to accelerate learning of Self-Organizing Maps(SOM) because there are many loops which are simple calculation in learning algorithm of SOM. Some approaches which are acceleration of learning of SOM by GPGPU have been proposed. In contrast to those proposals, our proposals are that some graphic processors calculate concertedly using "NVIDIA Scalable Link Interface technology(NVIDIA SLI)." In the experiments, we compare the learning speed and recognition rate of character recognition using SOMs which are calculated by MPU, a GPU(CUDA-SOM) and multi-GPU(SLI-SOM). As a result of experiments, the recognition rates for test patterns are almost identical in 3 approaches of SOM. In comparison of speed of learning, CUDA-SOM is about 7 times as fast as that in MPU in case of small size of competitive layer, and SLI-SOM about 80 times faster than MPU in large size of competitive layer. We succeeded in accelerating the learning performance from those results.

Keywords: Self-Organizing Maps, General Purpose on Graphics Processing Unit(GPGPU), CUDA, NVIDIA SLI, Character Recognition.

1 Introduction

The personal computers came to be able to calculate refined computation that are calculated by high performance computers until then because the performance of Micro Processing Units(MPU) increased significantly in the early 2000's. However, the performance improvement of it slowed down in the latter 2000's. The major factor of slowing is that semiconductor micro-fabrication is close to the limitation. This factor prevent the clock improvement of MPU, and this led to "Multi-Core/Many-Core" of Intel Corp. and AMD Inc.

S.-Y. Rhee et al. (eds.), *Soft Computing in Machine Learning*,
Advances in Intelligent Systems and Computing 273,
DOI: 10.1007/978-3-319-05533-6_5,

Unlike MPU, the graphics processing unit(GPU) have been used in ancillary, and GPUs are for acceleration of processing of 2 and 3 dimensional graphics generally. Recently, GPUs have been attempted to process for general computation, it are called "General Purpose Graphic on Processing Units(GPGPU)" and is the focus of attention[1]. The feature of GPU is that small and limited capable processors are parallelized on a massive scale, and those processors operate synchronously. GPGPU could be said that cores of MPU are simplified and increased large-scale. For this, GPGPU is similar of "Multi-Core/Many-Core" of MPU. Small processors in GPU, however, do not have general versatility same as MPU. For this reason, GPGPU have capability of acceleration of the specific purpose computation. The capability of GPGPU is specifically suitable for acceleration of simply calculation of loop iteration because processing is in parallel.

For example, Igarashi et al. proposed to simulate basal ganglia in real time by GPGPU[2], and James et al. adopt GPGPU to accelerate the "deep learning." [3]

In our research, we adopt GPGPU to accelerate learning of Self-Organizing Maps(SOM)[4]. There are many loops which are simple calculation in learning algorithm of SOM. Our proposal derive that loops from SOM algorithm, and it are processed by GPGPU in parallel. Also, sections which are not able to parallelize in algorithm of SOM are processed by MPU. Our proposal is able to expect to process faster than using MPU only, because it combines area of specialty of MPU and GPU as above. The "CUDA" which is developed by NVIDIA and the "ATI Stream" which is developed by ATI are available for GPGPU in some graphic processing units. The approaches which are acceleration of learning of SOM by GPGPU have been some proposed. Luo et al proposed to improve learning speed of SOM by CUDA[5] and ATI Stream, and Raghavendra is as same proposal[6].

In contrast to those proposals, the proposal in this study is that some graphic processors calculate concertedly the algorithm of SOM. We use the CUDA for acceleration of SOM, and the CUDA has a capability which operate with some GPUs concertedly called "NVIDIA Scalable Link Interface technology(NVIDIA SLI)." NVIDIA SLI enables some GPUs to share workload. There are 3 types of the SLI, i.e., "SLI", "Quad SLI" and "3-Way SLI." The SLI uses 2 graphics boards which are equipped a GPU. Not only the SLI but also the Quad SLI uses 2 graphics boards. The difference with the SLI is that graphics boards are equipped dual GPUs. On the other hand, the 3-Way SLI uses three graphics boards which are equipped a GPU. The graphics board which has 2 GPUs for the Quad SLI is so expensive. We use the 3-Way SLI to accelerate of SOM because it is excellent balance of cost-effectiveness.

In the experiments, we adopt SOM to character recognition. The experiments look into the effect of SLI on rate of recognition and difference of learning time. From experimental results, the SLI is 7 times in small competitive layer and the 3-Way SLI is 80 times faster in large competitive layer than MPU rate of recognition is much the same.

The details of GPGPU and the CUDA will be presented in section 2. Section 3 presents Kohonen SOM. Section 4 explain acceleration of learning of SOM by GPGPU. Section 5 presents experimental results. Section 6 concludes the paper.

2 GPU, NVIDIA CUDA and NVIDIA SLI

2.1 Graphic Processing Unit(GPU)

GPU have been originally dedicated processor for image processing, and, it has been used supplementarily under MPU for processing of time-varying images. Although early GPUs were for drawing 2D images, it has been used to draw fast for 3D images since around the 2000s. To draw 3D images, the operations which are simple but high parallelism are required. The manufacturer of GPU introduced the concept of thread into GPU for fast processing of those operations. In the mid-2000s, the length of bits in GPU increased to generate high precision 3D images. Therefore, GPU came to be able to carry out double-precision floating point arithmetic in recent years.

Owing to the improving of capability of GPU, it came to be used for not only processing of images but also general-purpose computation. Those of general-purpose computation by GPU are called "general-purpose computing on graphics processing unit(GPGPU)." The "CUDA" which is developed by NVIDIA and The "ATI Stream" which is developed by ATI are available for GPGPU in several graphic processing units. We use the "CUDA" which is developed by NVIDIA for our study, and the CUDA has a capability which operate with some GPUs concertedly called "NVIDIA Scalable Link Interface technology(NVIDIA SLI)."

2.2 The NVIDIA CUDA and the NVIDIA SLI

The CUDA. The basic architecture of the CUDA is as shown figure 1.

A GPU have many arithmetic units called "Streaming Multiprocessor(SM)", and there are many "CUDA cores" in a SM. A CUDA core is unit of minimum operation. CUDA cores belong to same SM execute same operations at a same time because a SM has arithmetic-logical units, registers and caches. Like SIMD, the operands are different although opcodes are same in an operation. For this, the speed of operations is accelerated. In case of processing by GPU, the data and operations have to be transferred to graphics memory on graphics boards from main memory because the graphics memory independent from main memory. After the calculation by GPU, the results are transferred to main memory from graphics memory. Before transfer, MPU issues an operation that synchronizes of CUDA cores to GPU. MPU is able to execute other operations while calculating by GPU. Because processing speed drastically falls, the frequency of data transfer have to be low. As mentioned above, many cores are arrayed in a GPU because the structure of GPU is simply. From these differences of structure, although memory bandwidth of MPU is about 40GB per second, that of GPU is about 150GB per second. Acceleration for calculation of floating-point

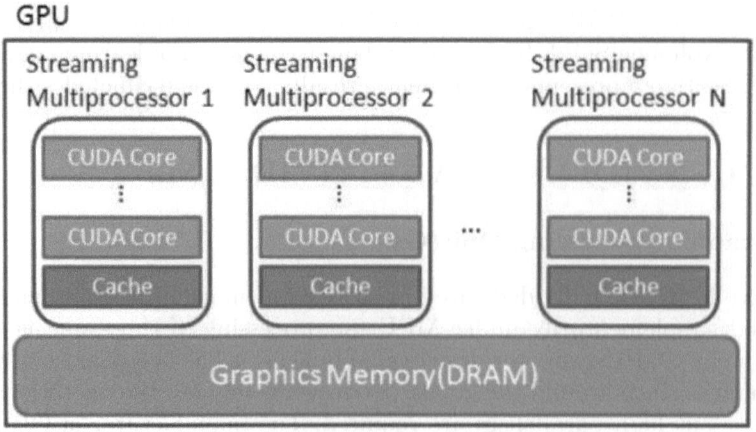

Fig. 1. The basic architecture of the CUDA

number is facilitated from advantage of memory bandwidth of GPUs. On the other hand, the conditional branching extremely slows the calculation because SMs do not have a branch predictor and units of Out-of-Order execution. Our study aims to accelerate the parts which are a paucity of conditional branching and are able to calculate independently in algorithm of SOM.

NVIDIA SLI Technology. The NVIDIA SLI Technology(SLI) is a technology to link some graphics boards which have a GPU or two GPUs. The SLI enables some GPUs to share workload. There are three types of the SLI, i.e., "SLI", "Quad SLI" and "3-Way SLI". The SLI uses 2 graphics boards which are equipped a GPU. Not only the SLI but also the Quad SLI uses two graphics boards. The difference with the SLI is that graphics boards are equipped dual GPUs. On the other hand, the 3-Way SLI uses three graphics boards which are equipped a GPU. The graphics board which has two GPUs for the Quad SLI is so expensive. We use 3-Way SLI to accelerate of SOM because it is excellent balance of cost-effectiveness.

3 Kohonen's Self-Organizing Maps(SOM)

Self-Organizing Maps(SOM) is the one of the machine learning algorithm without supervised signals. It has two layers, i.e., input and competitive layer, and the layers are fully connected. As shown in the equation 1 and 2, each neuron in the competitive layer has weight vectors, **w**, that are same dimensional input vector with the input layer. The basic SOM have a 2 dimensional competitive layer. Figure 2 shows the architecture of basic SOM.

$$\mathbf{x} = [x_1, x_2, x_3, \cdots, x_n] \in R^n \tag{1}$$

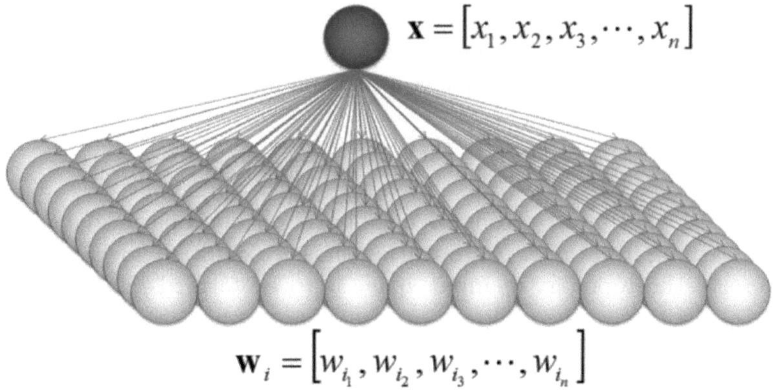

Fig. 2. The basic SOM

$$\mathbf{w}_i = [w_{i_1}, w_{i_2}, w_{i_3} \cdots, w_{i_n}] \in R^n \tag{2}$$

where $\mathbf{x}$ is a input vector, and $\mathbf{w}_i$ is a weight vector. i is the neuron number.

SOM algorithm adjusts in such a way that the distance norm between weight vectors of neurons in the competitive layer and input vectors in the input layer is small as follows. Firstly, weight vectors in neurons of the competitive layer are randomly initialized. Following this, neuron in the input layer is shown a vector of input pattern. Secondly, the distance norm between a certain input vector and a neuron is calculated, and the winner neuron which has the smallest distance is determined by equation 3 or 4.

$$c = \|\mathbf{x} - \mathbf{w}\| = \min_i\{\|\mathbf{x} - \mathbf{w}_i\|\} \tag{3}$$

$$c = \arg[\min_i\{\|\mathbf{x} - \mathbf{w}_i\|\} \tag{4}$$

In this study, the distance norm is Euclidian distance as,

$$\|\mathbf{x} - \mathbf{w}_i\| = \sqrt{\sum_j (x_j - w_{ij})^2} \tag{5}$$

Thirdly, the weight vectors in neighborhood neurons of the winner neuron are adjusted by equation 6.

$$\mathbf{w}_i(t+1) = \mathbf{w}_i(t) + h_{ci}(t)[\mathbf{x}(t) - \mathbf{w}_i(t)] \tag{6}$$

where t is time steps, and $h_{ci}(t)$ is step function in equation 7 or Gaussian function in equation 8.

$$h_{ct}(t) = \begin{cases} \alpha(t) : i \in N_c \\ 0 : i \notin N_c \end{cases} \tag{7}$$

$$h_{ct}(t) = \alpha(t) \cdot \exp\left(-\frac{\|\mathbf{r}_c - \mathbf{r}_i\|^2}{2\sigma_\Lambda(t)^2}\right) \tag{8}$$

where $\alpha(t)$, which is from 0 to 1, is a monotonically decreasing function for learning rate, $\mathbf{r}_c$ and $\mathbf{r}_i$ are position vectors of neuron c and i, N_c is set of neighborhood neurons of the winner neuron, and $\sigma_\Lambda(t)$ is as equation 9

$$\sigma_\Lambda(t) = (N_i - 1) \cdot \left(1 - \frac{t}{T_{\max}}\right) + 1 \tag{9}$$

where $N_i - 1$ is the initial radius of neighborhood neurons, and $t_{\max}$ is total number of learning iterations. By repeating those calculation several times, map which preserve the topology of input space is to be formed in the competitive layer.

4 CUDA-SOM and SLI-SOM

In this study, we propose to optimize the program of SOM for using a GPU or multi-GPUs. We call "CUDA-SOM" for using a GPU, and "SLI-SOM" for using multi-GPUs. As mentioned in Section 2.2, the conditions for optimization are that there are few conditional branches and that parallelization is possible.

In conventional approach, the part which calculate of Euclidean distances between input vectors and weight vectors in neurons is sequentially repeated computation, e.g., the for statement. That part is parallelizable because that distances are calculated in respective neurons, and there are no conditional branch in that calculation.

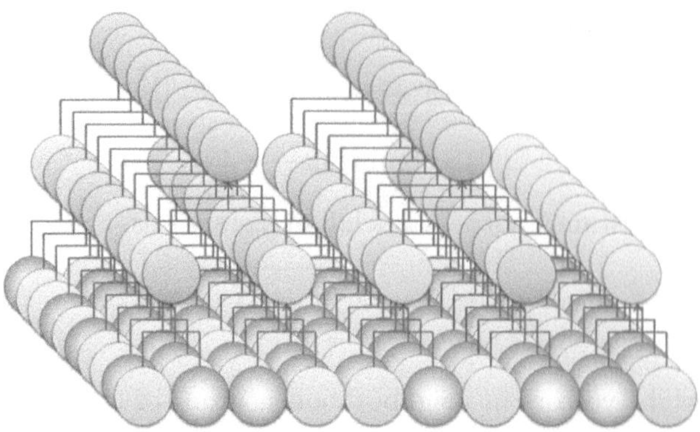

Fig. 3. The divide-and-conquer method. The blue and red neurons stand for winner between two neighbor.

In contrast to MPU, it uses divide-and-conquer method when the winner neuron is determined by GPU. Specifically, this part compares simultaneously Euclidean distances of two competitive neurons in parallel; afterward it selects neurons of small distance. These operations are performed repeatedly in Figure 3. This method has an advantage that there is no need to transfer data between main memory and graphics memory. The operations of data transfer slow computational speed. However, the determination of parameters for GPUs is difficult, and the program get complex. Furthermore, the processing may become not much fast because this method uses the if statement. The part which update the weight vectors in neurons of competitive layer is same as the part of calculation of Euclidean distances between input vectors and weight vectors because that updating are calculated in respective neurons. So, parallelization of this part is possible.

In "SLI-SOM", the competitive layer is split in 3 part for using the 3-Way SLI, and GPUs calculate above processes concertedly. Because the 3-Way SLI is able to manipulate a lot of threads which are units of processing more than single GPU, the parameters, e.g. the number of threads for a GPU, have to be made a decision carefully. Figure 4(a) shows the flow of acceleration of SOM algorithm using "CUDA-SOM" , (b) shows "SLI-SOM", and Table 1 shows the algorithm of "SLI-SOM."

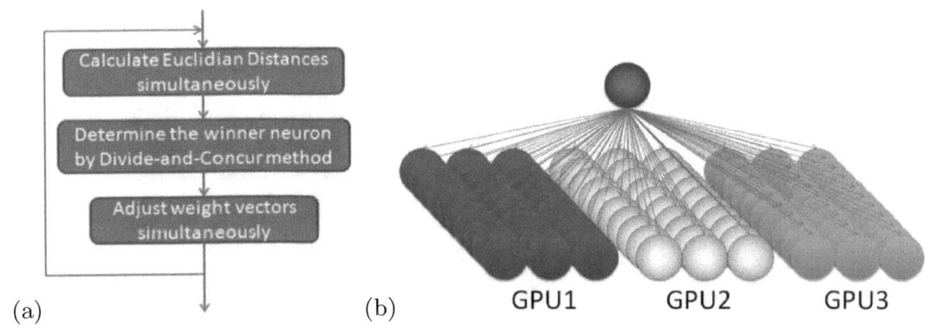

Fig. 4. CUDA-SOM and SLI-SOM. (a)Flow of acceleration. (b)How to use GPUs for SLI-SOM.

Table 1. The algorithm of SLI-SOM

	The algorithm of SLI-SOM learning
1	Read the input patter n
2	Calculate the Euclidian Distances between reference vectors and an input vector in simultaneously.
	GPU1 : 1/3 of competitive layer
	GPU2: 1/3 of competitive layer
	GPU3: 1/3 of competitive layer
3	Synchronize the information between GPUs
4	Determine the winner neurons in each GPUs
5	Transfer the information of the winner neurons in each GPUs
6	CPU determines the unique winner neuron

5 Experiments

5.1 Experimental Conditions

In this study, we use a computer which has "Intel Core i7 960@3.20GHz", 12G Bytes main memory, and three "NVIDIA GeForce GTX580 with 3.00G Bytes graphics memory" graphics boards. The experiments are pattern recognition. The SOM have 2 dimensional competitive layer, and the learning and test patterns for SOM are shapes of alphabet as shown in the Figure 5(a) and (b). There are 100 sets of test patterns. The shapes are discretized into 7 x 7 cells. The patterns are 49-dimensional vectors, and the black and white cells represent 1 and 0, respectively.

Fig. 5. Recognition patterns. (a)Learning patterns. (b)Example of test patterns.

Let the initial radius of neighborhood be same as one side of length of competitive layer, that of final be 1, the learning rate be 0.1, and learning epochs be 5000.

To determine the class label of neurons, the GPUs calculate simultaneously Euclidean distances between weight vectors in neurons and learning pattern. Subsequently, the GPUs calculate between weight vectors in neurons and test patterns in same manner for character recognition.

5.2 Experimental Results

Table 2 indicates that the recognition rates for learning and test patterns. That of MPU is lower than GPU in case of 10 square size of competitive layer. On the other hand, the recognition rates for learning patterns are 100% in the other size of competitive layer. The recognition rates for test patterns are almost identical in 3 methods. This result shows that MPU is possible to substitute in GPU about recognition rate.

Table 2. Comparison of recognition rate. The results are averaged over 3 times of recognition. The units is in percent.

Calculation method	Recognition pattern	Size of competitive layer						
		[10,10]	[20,20]	[30,30]	[100,100]	[150,150]	[200,200]	[300,300]
MPU	learning	98.72	100.00	100.00	100.00	100.00	100.00	100.00
	test	85.67	90.67	90.00	93.00	94.67	94.00	94.67
CUDA-SOM	learning	100.00	100.00	100.00	100.00	100.00	100.00	100.00
	test	90.00	87.67	90.00	93.00	94.33	91.33	93.67
SLI-SOM	learning	100.00	100.00	100.00	100.00	100.00	100.00	100.00
	test	88.00	89.33	89.00	92.67	93.00	92.00	94.00

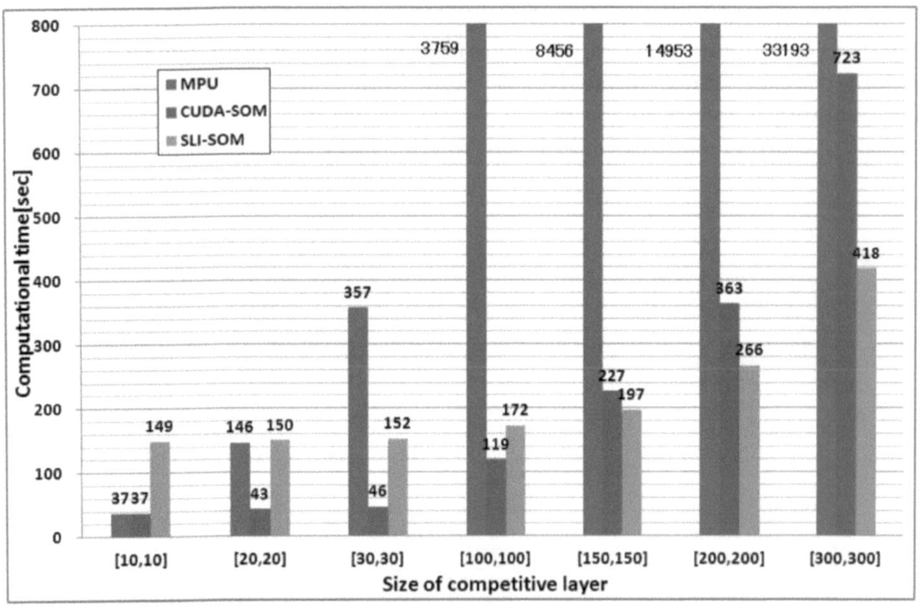

Fig. 6. Comparison of computational time. Upper and side numbers of bar graph stand for computational time, and it are in seconds.

Figure 6 shows the comparison of computational time between 3 methods. In case of MPU, the graphs which are over 100 square size of competitive layer are broken off because MPU method takes extremely long to calculate compared with case of GPU. Upper and side numbers of bar graph are computational time.

This result shows that CUDA-SOM is same as MPU when the size of competitive layer is 10 square. However, SLI-SOM is about 3 times slower than those proposal because SLI-SOM needs many data transfer. In 20 and 30 square size of competitive layer, computational time of CUDA-SOM and SLI-SOM increase very little. In contrast, that of MPU increases about 4 and 9 times, respectively.

In 100 square, CUDA-SOM increase about 3 times compared with former results. Though, SLI-SOM increases about 1.13 times. The computational time needs over 1 hour using MPU. The SLI-SOM is comparable with CUDA-SOM because data transfer and computational cost are balanced in case of 150 square. In 200 squares, the computational time is inverted between CUDA-SOM and SLI-SOM, and SLI-SOM is about 100 seconds faster than CUDA-SOM. In 300 square, the computational times of MPU needs about 9 hours. By contrast, that of CUDA-SOM needs about 12 minutes, and that of SLI-SOM needs about 7 minutes only. From the increasing rate of computational time, it is considered that time difference about larger size of competitive layer widen drastically because to share of workload becomes more important than data transfer.

The reason that CUDA-SOM has superior to SLI-SOM in small size of competitive layer is transfer overhead between main memory and graphics boards. Figure 7 shows the difference of CUDA-SOM and SLI-SOM. CUDA-SOM needs only one time of data transfer, on the other hand, SLI-SOM needs several times of data transfer. As mentioned in Section 4, the operations of data transfer slow computational speed. From this, CUDA-SOM is suitable for small size of competitive layer, and SLI-SOM is suitable for large size of competitive layer.

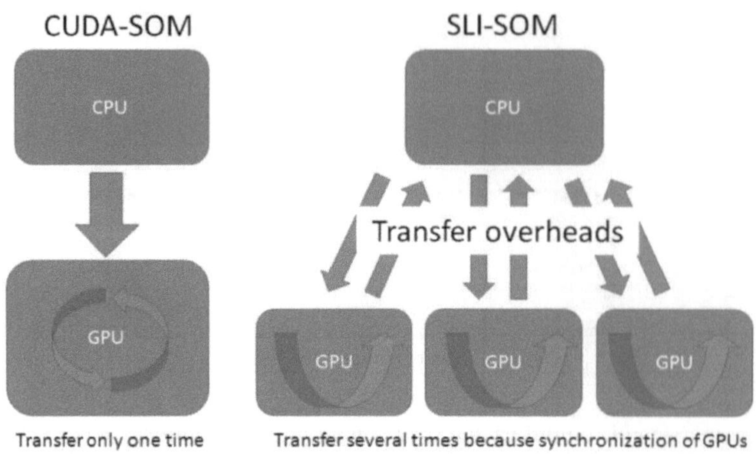

Fig. 7. Comparison of data transfer times between CUDA-SOM and SLI-SOM

6 Conclusions

In this paper, we have proposed to adopt GPGPU in order to accelerate the learning speed of SOM. Our key idea is that the multiple loops are resolved to many threads and it is calculated simultaneously by GPUs. The NVIDIA CUDA which is one of the GPGPU method is able to calculate by multi-GPU architecture, called the SLI. The SLI enables some GPUs to share workload.

In the experiments, we compare the learning speed and recognition rate of character recognition using SOMs which are calculated by MPU, a GPU(CUDA-SOM) and multi-GPU(SLI-SOM). The results show that the recognition rate is almost same between 3 methods. In computational time, MPU method increase linearly. In contrast, CUDA-SOM and SLI-SOM increase very little when the size of competitive layer is small, and CUDA-SOM is the fastest because data transfer between main memory and graphics memory is light. In case of large size of competitive layer, SLI-SOM is the fastest because to share of workload becomes more important than data transfer. As a result, we succeeded in accelerating the learning speed of SOM about 80 times faster than MPU using multi-GPU method.

Acknowledgments. This research was supported by Grant-in-Aid for Special Research Equipments from Nishinippon Institute of Technology, Japan.

References

1. Luebke, D., et al.: GPGPU: general-purpose computation on graphics hardware. In: Proc. of the 2006 ACM/IEEE conference on Supercomputing, Article No. 208 (2006)
2. Igarashi, J., et al.: Real-time simulation of a spiking neural network model of the basal ganglia circuitry using general purpose computing on graphics processing units. Neural Networks 24(9), 950–960 (2011)
3. Bergstra, J., et al.: Deep Learning on GPUs with Theano. The Learning Workshop - Research Abstract - Oral preferred (February 18, 2010)
4. Kohonen, T.: Essentials of the self-organizing map. Neural Networks 37, 52–65 (2013)
5. Zhongwen, L., et al.: Self-Organizing Maps Computing on Graphic Process Unit. In: Proc. of ESANN 2005, pp. 557–562 (2005)
6. Prabhu, R.D.: SOMGPU: An Unsupervised Pattern Classifier on Graphical Processing Unit. In: 2008 IEEE CEC, pp. 1011–1018 (2008)

Analysis of Local Concerts Using Facebook Adapting the Mathematical Model of Hit Phenomena

Yasuko Kawahata[1,*], Etsuo Genda[1], Chinami Hara[2], and Akira Ishii[2]

[1] Department of Contents and Creative Design, Graduate School of Design, Kyushu University
6-10-1 Hakozaki, Higashi-ku, Fukuoka 812-8581, Japan
purplemukadesan@gmail.com
[2] Department of Applied Mathematics and PhysicsTottori University,
Koyama, Tottori 680-8554, Japan
ishii.akira.t@gmail.com

Abstract. A mathematical model for the hit phenomenon in entertainment within a society is presented as a stochastic process of interactions of human dynamics. The calculations for the Japanese motion picture market based on to the mathematical model agree very well with the actual residue distribution in time. Views of Facebook could be predicted by the number of all the action in the viewer Facebook on this paper. In other words, Views So if determined by the actions for which you want to advertise, we will begin to see is how to should be utilized in advertising using Facebook future.

Keywords: Hit phenomena, Stochastic process, Advertisement, Facebook.

1 Introduction

Today, festivals centered around the local artists to activate the area actively have been made in Japan.[1,2] However, it has a challenge should carry out activities without imposing how the advertising costs in the process of the announcement of the activity. In this study, we targeted the troupe Tinker Bell in Fukuoka city. It is a theater company that men and women of all ages in their 50s from the 5-year-olds are participating. Mission of the troupe, to help you through the creative activities of on stage, to discover the existence value of each person team members, grow a heart and gentle rich sensitivity, how to improve the life of the team members that including children, it has set and to contribute to the civic culture by the creative musical quality by amateur and performance.

Without having made a public relations via the Internet so far, the first promotion for the theater company started in 2012 for this reserch. We thought doing web publicity for the theater company will also lead to stimulate motivation in the creation of the artist, and also help in subsequent activities. We can in the course in the activity of this time, Advertise the activities of the troupe to people living in other provinces in

* Corresponding author.

S.-Y. Rhee et al. (eds.), *Soft Computing in Machine Learning*,
Advances in Intelligent Systems and Computing 273,
DOI: 10.1007/978-3-319-05533-6_6, © Springer International Publishing Switzerland 2014

Fig. 1. Troupe Tinker Bell in Fukuoka

each announcement. In this method, we have measured the access number of official page on Facebook troupe. The number of pages of view for the theater company will tend to increase when a single event was held. We thought these increasing phenomenon is the hit phenomenon for the theater company. So we thought the application of mathematical models of the hit phenomenon this cases. The mathematical model of the hit phenomenon is possible to adapt to predict the number of page views of Facebook from the analysis result, we investigated in order to go to tips for us to go to public relations with less advertising expenses.

2 Hit Phenomena

Hit phenomenon is a phenomenon in which decays rapidly past the peak attendance and number of sales continue to increase at an explosive at the same time as the performance or sale by content such as music and movies get the support of the recipient. In other words, hit phenomenon is that it will spread to the public by petting its contents, such as speaking, listening and see the person-to-person.The fact they can be moved how people to "heart" In order to hit the various contents is important.

Thinking capture the movement of people in the course of hits happen in the "heart", how to continue to approach, is important in analyzing the hit phenomenon. We analysed that how will lead to hit phenomenon of interest in its content if allowed to approach people in the "heart" from the changes in the number of accesses generated between the content by applying conventional techniques in this paper.[3-7]

3 Mathematical Theory for Hit Phenomena for Facebook

The mathematical model for hit phenomena where the movement of human minds are considered as stochastic processes influenced by the media advertisement, communications with their friends and the rumor in the society. In this paper, the access with friends is considered as direct access or two-body interaction. The page in Facebook is considered as indirect access or three-body interaction. In this model, we define the intention for person "i" in the Facebook to activities as $I_i(t)$. The model can describe the intention of each person for viewing as the equation of the intention of person i with two-body interaction and three-body interaction terms. According to the Reference, we write down the equation of the intention at the individual level as

$$\frac{dI_i(t)}{dt} = -aI_i(t) + \sum_j d_{ij} I_j(t) + \sum_j \sum_k h_{ijk} d_{jk} I_j(t) I_k(t) + f_i(t) \tag{1}$$

where d_{ij}, h_{ijk}, and $f_i(t)$ are the coefficient of the direct access, the coefficient of the indirect access, and the random effect for person i, respectively. We consider the above equation for every person in the Facebook so that $i = 1$, N_p where N_p is the effective population of the Facebook for a certain Facebook pages.

The advertisement and publicity effect for each person can be described as the random effect $f_i(t)$.

Eq. (1) is the equation for all individual persons, but it is not convenient for analysis. Thus, we consider here the ensemble average of the intention of individual persons as follows:

$$\langle I(t) \rangle = \frac{1}{N} \sum_i I_i(t) \tag{2}$$

Taking the ensemble average of Eq. (1), we obtain for the left-hand side:

$$\left\langle \frac{dI_i(t)}{dt} \right\rangle = \frac{1}{N} \sum_i \frac{dI_i(t)}{dt} = \frac{d}{dt}\left(\frac{1}{N} \sum_i I_i(t) \right) = \frac{d\langle I \rangle}{dt} \tag{3}$$

For the right-hand side, the ensemble average of the first, second, and third is as follows:

$$\begin{aligned} \left\langle \sum_j \sum_k p_{ijk} I_j(t) I_k(t) \right\rangle &= \left\langle p \sum_j \sum_k I_j(t) I_k(t) \right\rangle \\ &= \frac{1}{N} \sum_i p \sum_j \sum_k I_j(t) I_k(t) \\ &= \sum_i p \frac{1}{N} \sum_j \sum_k I_j(t) I_k(t) \\ &= Np \sum_i \frac{1}{N} \sum_j I_j(t) \frac{1}{N} \sum_k I_k(t) \\ &= N^2 p \langle I(t) \rangle^2 \end{aligned} \tag{4}$$

$$\langle -aI_i \rangle = -a \frac{1}{N} \sum_i I_i(t) = -a \langle I(t) \rangle$$

$$\left\langle \sum_j d_{ij} I_j(t) \right\rangle = \left\langle \sum_j dI_j(t) \right\rangle = \frac{1}{N} \sum_i \sum_j dI_j(t) = \sum_i d \frac{1}{N} \sum_j I_j(t) = Nd \langle I(t) \rangle \tag{5}$$

where we assume that the coefficient of the direct and indirect access can be approximated to be under the ensemble average.

For the fourth term of Eq. (1), the random effect term, we consider that the random effect can be divided into two parts: the collective effect and the individual effect:

$$f_i(t) = \langle f(t) \rangle + \Delta f_i(t) \tag{7}$$

Facebook Action are the very important factor to increase the intention of each person for viewing pages in Facebook. We consider the action effect as an external force term A(t) to the intention as follows,

$$\frac{dI_i(t)}{dt} = -aI_i(t) + A(t) + \text{access} \tag{8}$$

where $A(t)$ is the Facebook's advertisement effect. The actual formula used in the calculation to analyze the SNS response are introduced in detail in Reference.[7-10]

Table 1. Parameta List

alpha	Interest willingness attenuation ratio of daily
Cadv	Percentage of increase of Views from behavior of visitors
before	Cadv to become stronger as it approaches the peak of access number
after	Cadv to become weaker with increasing distance from the peak of access number
a	Percentage of Cadv attenuation of each course of the day
NpDnn	Direct access (Between people have not seen the article or content)
NpDny	Direct access (People who have not seen a person who saw the content and articles)
Np2Pnn	Indirect access (between people have not seen the content articles and touch the content)
Np2Pny	Indirect access (people you have not seen a person who looked at content articles and touch)

4 Results

We chose the timing of the two events in order to measure the number of access Facebook page to be analyzed. We thought that Views can be predicted from visitors total number of actions period of two and useful publicity theater company in the future.

① We exhibited six points to short movie contest of Misuzu Kaneko a Movie of musical theater company in October 2012. [11,12]

② We held workshop by professional actor who is active in the world like an broadway. So we have announced on Facebook in March 10 2013. At this timing, the Web page of the troupe also gathered many visitors number most from people who did not know. [13]

③ We held public presentation of "Kaneko misuzu of World" in April 14 2013 at Hakata citizen's Center. At this timing, the Web page of the troupe also gathered many visitors number most from people who did not know. We have also analyzed in order to investigate indirect access or shown what parameters. [14]

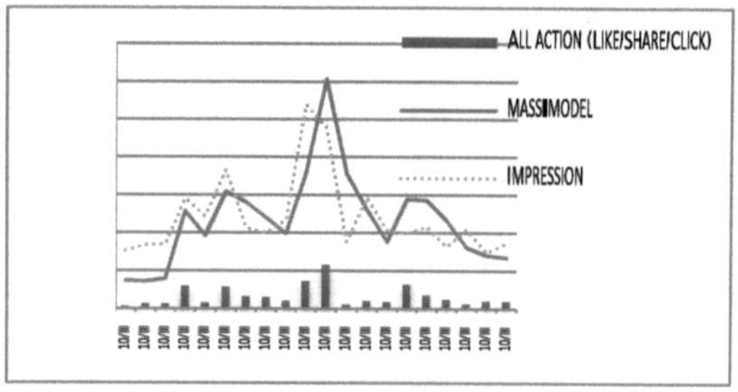

Fig. 2. Short movie contest of Misuzu Kaneko a Movie of musical theater company in October 2012

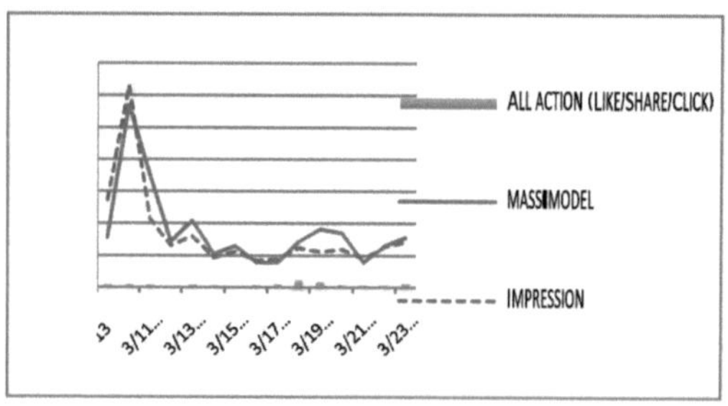

Fig. 3. Actor playing an active part in the world of professional, such as a Broadway workshop of theater company presided over came as a guest lecturer. in March 10 2013

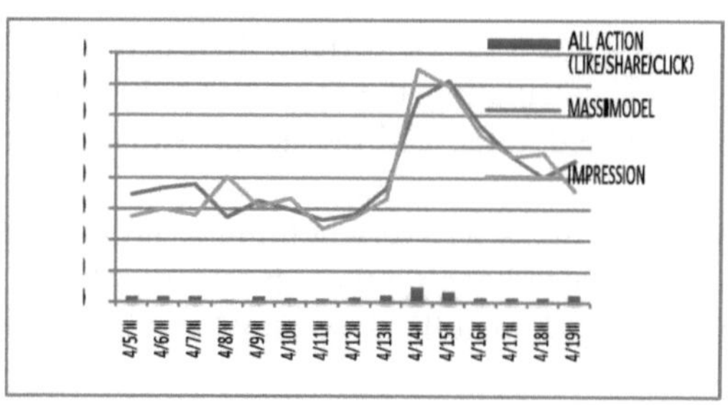

Fig. 4. We held public presentation of "Kaneko misuzu of World" in April 14 2013 at Hakata citizen's Center

5 Conclusion

We found the following four points in this study. We are a comparison of the parameter in each period of the two activities that target this result data is shown below.

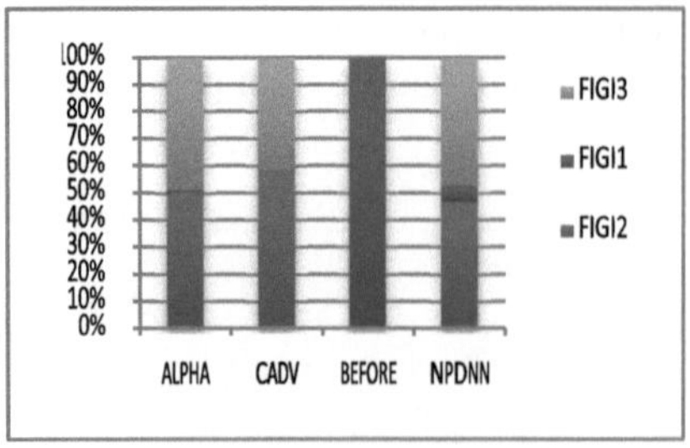

Fig. 5. Comparison chart of the parameters (alpha, Cadv, before, a)

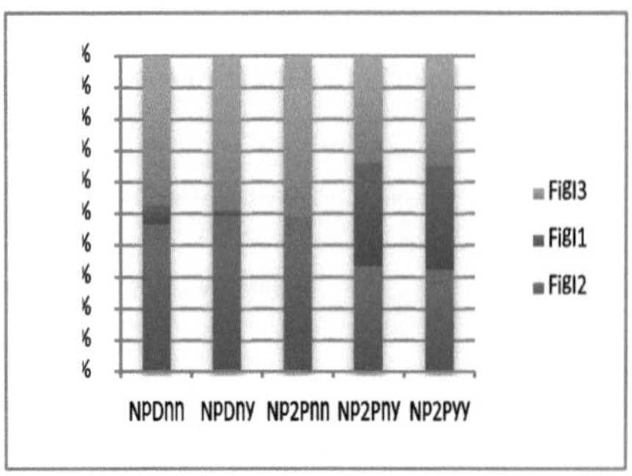

Fig. 6. Comparison chart of the parameters (NpDnn, NpDny, Np2Pnn, Np2Pny, Np2Pyy)

We found the following four points in this study.

① Number of all actions as a factor variable that determines .Number of prediction have increased by calculating the factor variable.However, there are some differences in the predicted value and the actual value if the impact of the event itself is large.

② Number of views of Fig.3 was overwhelmingly proportion of Cadv.

After the workshop was also keeping the advertising effect. Attention to the professional actor in the workshop at the time may have led to this result.

③ We need to do the analysis and application of this model with respect to advertisement methods of professional artists in the middle of going to actively publicity by the media such as TV and magazines as a policy of future research. Future, empirical research will be with respect to parameters of indirect access and direct access by making a comparison with the organic number of users who access the actual.

Acknowledgement. The authors would like to thank CEO Uchiyama Kouki of Hotlink, CEO Koguchi Hidehiko of Perspective Media , Theatre TinkerBell for giving them Database and helpful discussions. Great creaters who have supported the motivation of this study, We also thank Troup Tinker Bell in Fukuoka (Hara Chinami and Kinoshita Tsuguo, Hisada Shigeki) and versity KANSEI center for Arts and Science, ADAA (Asia Digital Arts Awards) which gives the best environment to do their studies in particular, to carry out development activities creativity of Asia. In adidition , it is enormously grateful family and friends and Genda Lab. Members , also ancestors who have supported much emotional support. Y.Kawahata also thanks the Japan Student Services Organization for the scholarship.

References

1. Kawahata, Y.: Use of SNS as a method of outreach. In: Japan Association for Arts Management of Kyushu, Workshop Report Collection for Students (2012)
2. Theatre Tinker Bell, `https://www.facebook.com/JuTuantinkaberu`
3. Ishii, A., Arakaki, H., Matsuda, N., Umemura, S., Urushidani, T., Yamagata, N., Yoshida, N.: New Journal of Physics 14, 063018 (22pp) (2012)
4. Ishii, A., Matsumoto, T., Miki, S.: Prog. Theor. Phys. (suppl. 194), 64–72
5. Kawahata, Y., Genda, E., Ishii, A.: Revenue Prediction of Music Concerts Using the Mathematical Model of Hit Phenomena. In: ICBAKE 2013 (2013) (unpublished)
6. Ishiil, A., Ota, S., Koguchi, H., Uchiyama, K.: Quantitative analysis of social popularity for Japanase pop girl group AKB48 using mathematical model for hit phenomena. In: ICBAKE 2013 (2013) (unpublished)
7. Ishii, A., Furuta, K., Oka, T., Koguchi, H., Uchiyama, K.: Mathematical Model of Hit Phenomena as a theory for collective motion of human mind in societies. In: The Frontiers of Artificial Intelligence and Applications (FAIA) Series. IOS Press (2012) (in press)
8. Ferebee, S.S., Davis, J.W.: Use of SNS as a method of outreach. In: Proceedings of the 4th International Conference on Persuasive Technology. ACM (2009)
9. Ross, C., et al.: Personality and motivations associated with Facebook use. Computers in Human Behavior 25(2), 578–586 (2009)
10. Viswanath, B., et al.: On the evolution of user interaction in facebook. In: Proceedings of the 2nd ACM Workshop on Online Social Networks. ACM (2009)
11. Kaneko Misuzu short movie contest, `https://www.facebook.com/misuzu.smc2012`
12. Misuzu 燦参 SUN 2012 (2012), `http://misuzu333.jp/`
13. Sawaki Jun Workshop, `http://tkbell.cocolog-nifty.com/blog/2013/03/post-6c0f.html` (retrieved)
14. We held public presentation of "Kaneko misuzu of World" in April 14 2013 at Hakata citizen's Center, `http://hakata-fukuoka.mypl.net/event/00000114698/` (retrieved)

Effective Selection and Mutation in Genetic Algorithm for PAPR Reduction of OFDM Signal

Noritaka Shigei, Kentaro Araki, and Hiromi Miyajima

Kagoshima University, 1-21-40 Korimoto, Kagoshima 890-0065, Japan
{shigei,miya}@eee.kagoshima-u.ac.jp, k2287401@kadai.jp

Abstract. Orthogonal frequency division multiplexing (OFDM) is superior in spectral efficiency and is widely used in today's digital communication. One of the drawbacks of OFDM is that the peak-to-average power ratio (PAPR) of the transmitted signal tends to be high. In order to overcome this problem, peak power reduction methods based on tone injection have been proposed. The peak power reduction problem solved with tone injection (TI) is a combinatorial problem. In this paper, we propose an improved genetic algorithm (GA) for the PAPR reduction based on TI. In order to find better solutions in a short time, improved selection and mutation schemes are proposed. The effectiveness of the GA method is demonstrated by numerical simulations in terms of PAPR, computation time and bit error rate (BER).

1 Introduction

Recent advance in digital signal processing technology demands faster wireless communication. Orthogonal frequency division multiplexing (OFDM) [1] is superior in spectral efficiency and is widely used in today's digital communication. One of the drawbacks of OFDM is that the peak-to-average power ratio (PAPR) of the transmitted signal tends to be high. In order to overcome this problem, various peak power reduction methods have been proposed[2–7]. Tone injection is an effective technique to overcome the PAPR problem[6, 7]. The peak power reduction problem solved with tone injection (TI) is a combinatorial problem. In [6], a greedy searching algorithm has been proposed. Its main drawback is to easily get a stack shallow local minimum. In [7], the solution space is reasonably reduced by introducing some constraints into the TI technique, and a neural network (NN) has been applied to the TI method. Further, in [8, 9], genetic algorithms (GAs) have been applied to the PAPR reduction method, it has been shown that the proposed GA methods provide the better performance than the NN method and random searches. However, the application to a digital communication system desires a more effective GA method in terms of both searching time and searching ability. Because this makes possible the enhancements of communication quality, bit rate and energy efficiency.

In this paper, we propose an improved GA algorithm for the PAPR reduction. In order to find better solutions in a short time, improved selection and mutation schemes are proposed. The selection scheme is based on not only a fitness

S.-Y. Rhee et al. (eds.), *Soft Computing in Machine Learning*,
Advances in Intelligent Systems and Computing 273,
DOI: 10.1007/978-3-319-05533-6_7, © Springer International Publishing Switzerland 2014

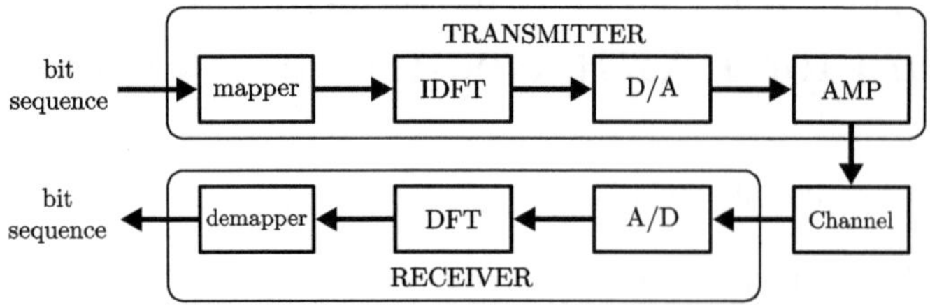

Fig. 1. Flow of OFDM from a Transmitting End to a Receiving End

function but also the ranking of the fitness values. The mutation scheme makes use of the information whether the current solution has already appeared or not in the past, and if that is the case, the scheme mutates the solution into a novel one with a high probability. The effectiveness of the GA method is demonstrated by numerical simulations in terms of PAPR, computation time and bit error rate (BER).

2 OFDM and Tone Injection

OFDM uses multiple subcarriers that are orthogonal to each other. Let T be the OFDM symbol time. The subcarriers are spaced $1/T$ Hz apart from each other. The flow of OFDM from a transmitting end to a receiving end is shown in Fig.1.

At the transmitter side, a bit sequence b_{L-1}, b_{L-2}, $\cdots$, b_0 to be transmitted is converted into a sequence of complex symbols X_0, X_1, $\cdots$, X_{N-1} by mapper. In this conversion, each m-bit subsequence in L-bit sequence is mapped to a complex number according to the used digital modulation scheme such as phase-shift keying (PSK) and quadrature amplitude modulation (QAM). When using M-QAM, X_n represents $\log_2 M$-bit subsequence $b_{L-m \cdot n-1}$, $b_{L-m \cdot n-2}$, $\cdots$, $b_{L-m \cdot n-m}$, where $L = N \log_2 M$ and $m = \log_2 M$. Samples of OFDM signal Y_0, Y_1, $\cdots$, Y_{N-1} are generated by IDFT (Inverse Discrete Fourier Transform) as follows:

$$Y_n = \frac{1}{N} \sum_{k=0}^{N-1} X_k e^{j \frac{2\pi}{N} nk}. \tag{1}$$

The IDFT operation can be performed by IFFT (Inverse Fast Fourier Transformation) in $O(N \log N)$ steps. The samples are converted to an analog signal, and then the signal is amplified and is fed to the transmission channel.

The demodulation process is performed in reverse order of transmitter's operations. From the received OFDM signal, N samples $\widetilde{Y}_0$, $\widetilde{Y}_1$, $\cdots$, $\widetilde{Y}_{N-1}$ are extracted with T/N sampling interval. Symbols $\widetilde{X}_0$, $\widetilde{X}_1$, $\cdots$, $\widetilde{X}_{N-1}$ are restored by DFT (Discrete Fourier Transform) as follows:

$$\widetilde{X}_n = \sum_{k=0}^{N-1} \widetilde{Y}_k e^{-j \frac{2\pi}{N} nk}. \tag{2}$$

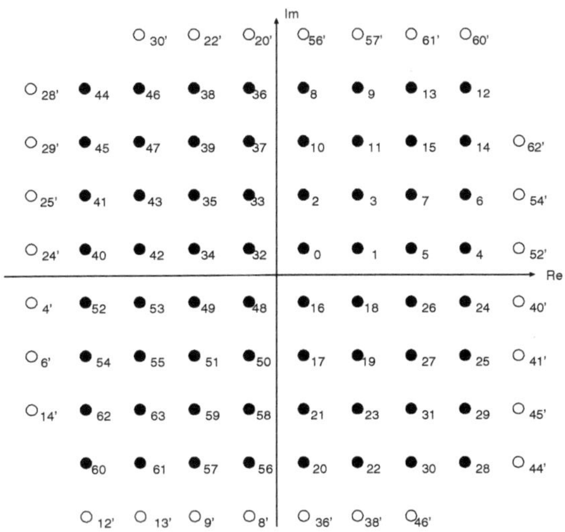

Fig. 2. The Used Constellation Map of 64-QAM

The restored symbol sequence $\widetilde{X}_0$, $\widetilde{X}_1$, $\cdots$, $\widetilde{X}_{N-1}$ is converted to a bit sequence $\tilde{b}_{L-1}, \tilde{b}_{L-2}, \cdots, \tilde{b}_0$. If $\widetilde{X}_n = X_n$ for all $n \in \{0, 1, \cdots, N-1\}$, the receiver retrieves the transmitted bit sequence b_{L-1}, b_{L-2}, $\cdots$, b_0 with no error, that is, for all $l \in \{0, 1, \cdots, L-1\}$, $\tilde{b}_l = b_l$.

The OFDM signal often has a very high peak power compared to its average power, because the signal is produced as a synthetic signal of a number of subcarriers. The degree of the peak power is evaluated by peak-to-average power ratio (PAPR) Φ defined as follows:

$$\Phi = \frac{\max_{0 \le k < N} |Y_k|^2}{E\left\{|Y_n|^2\right\}}, \tag{3}$$

where $E\{|Y_n|^2\}$ is the average power of the OFDM signal and $\max_{0 \le k < N} |Y_k|^2$ is the peak power of the OFDM signal.

Tone injection (TI) suppresses the peak power by relocating some complex symbols from their original position to other position[6, 7]. Assume that the used digital modulation scheme is M-QAM with an $\sqrt{M} \times \sqrt{M}$ constellation grid, and the minimum distance between constellation points is d. The original position of a complex symbol, which will be referred as original symbol, is located at a grid point of the $\sqrt{M} \times \sqrt{M}$ grid, and the destination of relocation is located outside of the grid.

Fig.2 shows the constellation map of 64-QAM used in [7], which will be used in this paper. In the figure, 64 black circles indicate the original symbols of 64-QAM and 28 white circles indicate the relocated symbols. The black circle with a number $m \in \{0, 1, \cdots, 63\}$ is the original symbol $X^{(m)}$, and the white circle with a number m' is the replacement of the original symbol $X^{(m)}$.

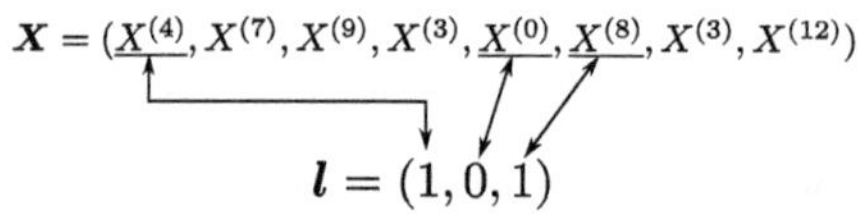

Fig. 3. An Example of Chromosome for $N = 8$ and $S = 3$

3 Conventional GA

We have applied GA to the peak power reduction method based on TI[8, 9]. This section explains the methods proposed in[8, 9], which use GA to find an effective combination of symbols to be moved for PAPR reduction.

3.1 Genetic Representation

In GA, a solution is coded as a string, called chromosome. Let $\mathcal{A} = \{X^{(0)}, X^{(1)}, \cdots, X^{(M-1)}\}$ be the set of all the original complex symbols. Let $\mathcal{M}$ be the set of the movable original complex symbols. That is, in Fig.2, the $\mathcal{A}$'s elements correspond to the black circles, and the $\mathcal{M}$'s elements correspond to the white circles. Let $\boldsymbol{X} = (X_0, X_1, \cdots, X_{N-1})$ be the vector of original complex symbols to be transmitted. Let $\mathcal{S} = \{X_n \in \mathcal{M} | 0 \leq n \leq N-1\}$ and $S = |\mathcal{S}|$ be the set of movable symbols in $\boldsymbol{X}$ and the size of $\mathcal{S}$, respectively.

A solution (chromosome) for $\boldsymbol{X}$ is coded as an S-dimensional binary vector

$$\boldsymbol{l} = (l_1, l_2, \cdots, l_S), \tag{4}$$

where $l_s \in \{0, 1\}$, and $l_s = 1$ and $l_s = 0$ mean that the symbol $X_{n(s)}$ is moved and is not moved, respectively, where $n(s)$ is the index of the s-th movable element in $\boldsymbol{X}$. In GA, l_s is called gene. Fig.3 shows an example of the chromosome coding. Gene l_s in $\boldsymbol{l}$ corresponds to $X_{n(s)}$.

3.2 Flow of GA

The algorithm of GA is as follows.

Step 1: Set the current generation $g \leftarrow 1$. Set the best PAPR $\Phi_{\text{best}} \leftarrow \infty$.
Step 2: Randomly generate K individuals $\boldsymbol{l}_1, \boldsymbol{l}_2, \cdots, \boldsymbol{l}_K$ in the form of Eq.(4).
Step 3: For each $k \in \{1, 2, \cdots, K\}$, search for the fitness f_k and PAPR Φ_k in the fitness table.

$$f_k = \left(\frac{1}{\Phi_k}\right)^{\alpha}, \tag{5}$$

where $\alpha > 0$ is a constant value, and where f_k and Φ_k are k-th individual's fitness and PAPR, respectively. The fitness table is implemented as a hash table, in which any item is searchable in a constant time. If f_k is not found, calculate Φ_k and f_k and register them into the fitness table, where the calculation requires $O(\log N)$ time.

Let $\Phi_{\min} = \min_{k\in\{1,2,\cdots,K\}} \Phi_k$. Let $k_{\min} \in \{1,2,\cdots,K\}$ such that $\Phi_{k_{\min}}$. If $\Phi_{\min} > \Phi_{\text{best}}$, then $\Phi_{\text{best}} \leftarrow \Phi_{\min}$ and $\boldsymbol{l}_{\text{best}} \leftarrow \boldsymbol{l}_{k_{\max}}$.
Step 4: If $g > G$, then go to Step 10.
Step 5: For each $k \in \{1,2,\cdots,K\}$, calculate the selection probability according to the selection probability

$$P_k = \frac{f_k}{\sum_{i\in\{1,2,\cdots,K\}} f_i}. \tag{6}$$

Draw K samples $\boldsymbol{l}'_1, \boldsymbol{l}'_2, \cdots, \boldsymbol{l}'_K$ with replacement from $\boldsymbol{l}_1, \boldsymbol{l}_2, \cdots, \boldsymbol{l}_K$ with probabilities $P_1, P_2, \cdots, P_K$.
Step 6: For each pair $\boldsymbol{l}'_{2k+1}$ and $\boldsymbol{l}'_{2k+2}$ $(0 \le k \le \frac{K}{2} - 1)$, perform the crossover operation with probability P_{c}.
Step 7: For each locus l_s in $\boldsymbol{l}'_1, \boldsymbol{l}'_2 \cdots \boldsymbol{l}'_K$, perform the mutation operation with probability P_{m}.
Step 8: Copy $\boldsymbol{l}'_1, \boldsymbol{l}'_2, \cdots, \boldsymbol{l}'_K$ to $\boldsymbol{l}_1, \boldsymbol{l}_2, \cdots, \boldsymbol{l}_K$, respectively.
Step 9: $g \leftarrow g + 1$ and go to Step 3.
Step 10: Return $\boldsymbol{l}_{\text{best}}$ as the final solution. □

3.3 Genetic Operators

In[9], four types of crossover operators, one-point, two-point, three-point and uniform, have been tested, and it has been shown that the uniform operator provides the best performance among them. Given a pair of individuals $\boldsymbol{l}_a$ and $\boldsymbol{l}_b$, the uniform operator randomly swaps genes, between $\boldsymbol{l}_a$ and $\boldsymbol{l}_b$.

The used mutation operator operates on each gene locus of chromosomes. Given a chromosome $\boldsymbol{l} = (l_1, l_2, \cdots, l_S)$, each gene locus $l \in \{l_1, l_2, \cdots, l_S\}$ is updated as follows:

$$l \longleftarrow \begin{cases} (l+1) \bmod 2 & \text{with probability } P_m \\ l & \text{with probability } 1 - P_m. \end{cases} \tag{7}$$

3.4 Fitness Table

In Step 3, the fitness table is employed for reducing the computation time. The evaluation of a solution involves IFFT operation, which requires $O(N \log N)$ steps. The hash-based implementation introduced in [9] makes it possible to search and insert an item in $O(1)$ steps.

4 Proposed GA

In this section, we present a selection scheme and a mutation operator.

4.1 Hybrid Selection

The conventional GA used a fitness proportionate selection, in which the selection probability is proportional to the fitness value as Eq.(6). However, when the deviation of the fitness values is too small, this selection cannot raise the proportion of relatively good individuals. On the other hand, the rank based selection determines the selection probability of an individual according to the ranking of fitness values. Although the rank based selection can overcome the difficulty for the fitness proportionate selection, the selection does not take into account the degree of the difference of fitness values.

In order to overcome the difficulties for both selection schemes, we propose a hybrid of the fitness proportionate selection and the rank based selection. The following Step 5.1∼ 5.3 is used in Step 5 of section 3.2 as the proposed selection.

Step 5.1: For each $k \in \{1, \cdots, K\}$, calculate the rank $r_k \in \{1, \cdots, K\}$ of the individual $\boldsymbol{l}_k$, where the fitness value f_k is r_k-th largest among $f_1, f_2, \cdots, f_K$.
Step 5.2: For each $k \in \{1, \cdots, K\}$, calculate the scaled fitness value $\tilde{f}_k$ as follows:

$$\tilde{f}_k = R(r_k) \cdot f_k, \tag{8}$$

where the scaling factor $R(r) \geq 0$ is a decreasing function of r. The specific form of $R(r)$ will be presented later.
Step 5.3: For each $k \in \{1, \cdots, K\}$, calculate the probability of selection P_k as follows:

$$P_k = \frac{\tilde{f}_k}{\sum_{i \in \{1,2,\cdots,K\}} \tilde{f}_i}. \tag{9}$$

Draw K samples $\boldsymbol{l}'_1, \boldsymbol{l}'_2, \cdots, \boldsymbol{l}'_K$ with replacement from $\boldsymbol{l}_1, \boldsymbol{l}_2, \cdots, \boldsymbol{l}_K$ with probabilities $P_1, P_2, \cdots, P_K$. □

The design of the scaling factor $R(r)$ is a key element in the proposed selection. The simplest function for $R(r)$ is a linear function such as $R(r) = -a \cdot r + b$ where $a > 0$, $b \geq 0$ and $R(r) \geq 0$. However, the linear function lacks flexibility. For example, the function cannot approximate a mix of smooth curve and steep curve. In this paper, we propose the following scaling factor function consisting of two linear functions.

$$R(r) = \begin{cases} \Gamma_1 - \dfrac{\Gamma_1 - 1}{\gamma - 1}(r-1) & \text{for } r \leq \gamma \\ \Gamma_K + \dfrac{1 - \Gamma_K}{K - \gamma}(K - r) & \text{for } r > \gamma, \end{cases} \tag{10}$$

where Γ_1 and Γ_K are the scaling factor for $r = 1$ and $r = K$, respectively, and where γ is r's value at the intersection of the two linear functions. Note that $R(r)$ with $\gamma = K$ is in the form of a linear function. Fig.4 shows two examples of $R(r)$. Eq.(10) involves three parameters Γ_1, Γ_K and γ to be determined. The effective parameter values will be determined by numerical simulation.

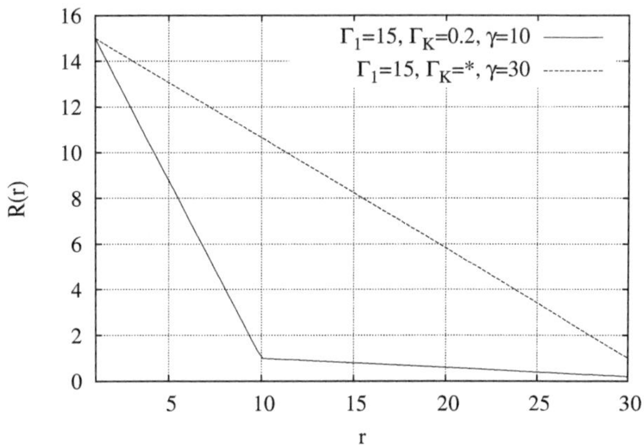

Fig. 4. Examples of $R(r)$ for $K = 30$

4.2 Mutation Based on Past Record

In a typical conventional GA, most of individuals in a population become similar as the generation progresses. The convergence of a population may be necessary property for finding a global optimal solution. From the other point of view, the later generations wast time, because the number of generated novel solutions decreases with the generation and the improvement of the solution becomes smaller with the generation. The key genetic operator for overcoming the difficulty is mutation. The conventional operator mutates each gene with a fixed probability as described in section 3.3. However, when the probability of mutation is small, the conventional operator fails to generate a novel solution with a high probability. In practical, the mutation probability should be small to obtain a better performance.

In order to produce a novel solution with a high probability in later generations, we propose a mutation operator based on the past record of generated chromosomes. The proposed mutation checks whether each chromosome generated by the crossover operation has already appeared or not, and if that is the case, the operator mutates the chromosome into a novel one with a high probability. Usually, it is a very time consuming process to check whether it is a novel or not. However, our conventional GA implements the fitness table by using a hash table. Therefore, the check process can be completed in a constant time.

The proposed mutation is introduced into the GA algorithm by modifying Steps 3 and 7 in section 3.2 as follows:

Step 3: For each $k \in \{1, 2, \cdots, K\}$, search for the fitness f_k and PAPR Φ_k in the fitness table. If f_k is found, a gene locus randomly selected is mutated and calculate Φ_k and f_k and register them into the fitness table.

Let $\Phi_{\min} = \min_{k\in\{1,2,\cdots,K\}} \Phi_k$. Let $k_{\min} \in \{1, 2, \cdots, K\}$ such that $\Phi_{k_{\min}}$. If $\Phi_{\min} > \Phi_{\text{best}}$, then $\Phi_{\text{best}} \leftarrow \Phi_{\min}$ and $\boldsymbol{l}_{\text{best}} \leftarrow \boldsymbol{l}_{k_{\max}}$.
Step 7: Do nothing. □

Table 1. The parameters used in the simulation

Method	N	α	Γ_1	Γ_K	γ	P_m
Proposed GA	128	8	200	0.1	10	–
	256	10	500	0.1	10	–
GA with proposed selection	128	3	2000	0.02	17	0.02
	256	4	1000	0.5	19	0.01
GA with proposed mutation	128	16	–	–	–	–
	256	34	–	–	–	–
Conventional GA	128	8	–	–	–	0.02
	256	12	–	–	–	0.01

5 Numerical Simulations

In the simulations, the following four types of methods are evaluated.

- **Proposed GA:** the method is described in sect. 4.
- **GA with proposed selection:** the method uses hybrid selection described in sect. 4.1 and conventional mutation described in sect. 3.3.
- **GA with proposed mutation:** the method uses conventional selection described in sect. 3.2 and proposed mutation described in sect. 4.2.
- **Conventional GA:** the method is described in sect. 3.

For every methods, the population size is $K = 30$, the maximum generation number is $G = 50$, and the crossover probability is $P_c = 0.9$. Table 1 shows the other parameters used in the simulation. These parameters are effective values found in preliminary experiments.

5.1 PAPR Reduction Performance

Fig. 5 shows the simulation results on PAPR versus the number of generations for $N = 128$ and $N = 256$, respectively. Every plotted points are averaging values for 2000 trials. At $G = 50$ for both $N = 128$ and $N = 256$, 1) the proposed GA achieves the minimum PAPR, 2) the GAs with either of proposed selection and proposed mutation achieves a better PAPR than the conventional one, and 3) the GA with proposed mutation achieves a better PAPR than the one with proposed selection. For both $N = 128$ and $N = 256$, two pairs, "Proposed GA and GA with proposed mutation" and "GA with proposed selection and Conventional GA", show smaller differences than other pairs. However, their differences are statistically significant. Their p-values of Welch's t test are shown in Table 2.

Fig. 6 shows the evaluation results on the normalized computation time versus the number of generations for the proposed GA and the conventional GA. The CPU time required for 300 runs is measured, and each plotted value is normalized by dividing it by the CPU time for the corresponding N and $G = 50$. The differences between proposed GA and conventional GA are obviously statistically significant. For $N = 128$ and $N = 256$, the computation time of the proposed GA is 1.62 and 1.41 times larger than the one of the conventional GA, respectively. This is because, the conventional GA skips the PAPR calculation for the solutions previously appeared, but the proposed mutation involves the PAPR calculation for every K individuals in a generation.

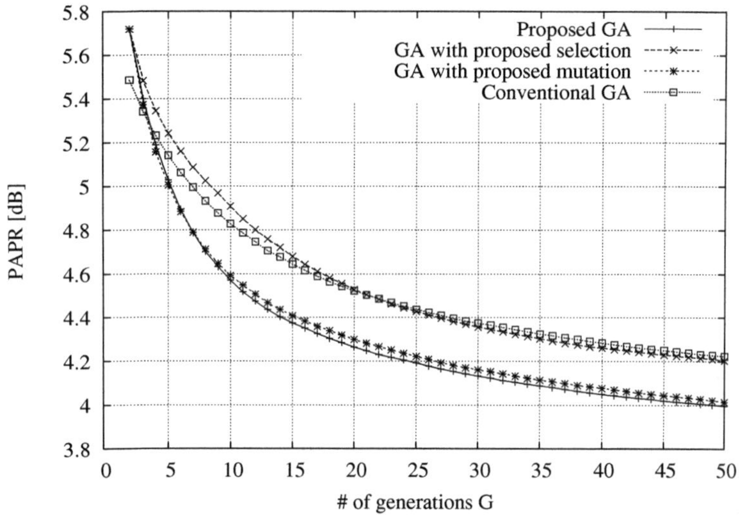

(a) For $N = 128$.

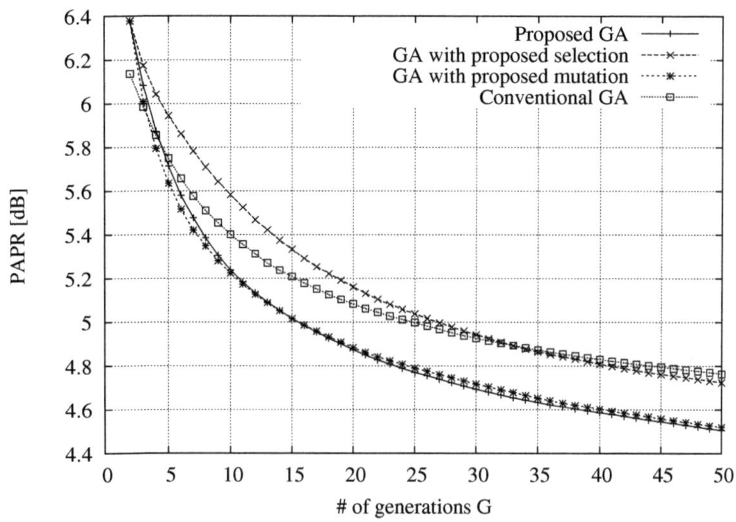

(b) For $N = 256$.

Fig. 5. PAPR versus the number of generations G for $N = 256$

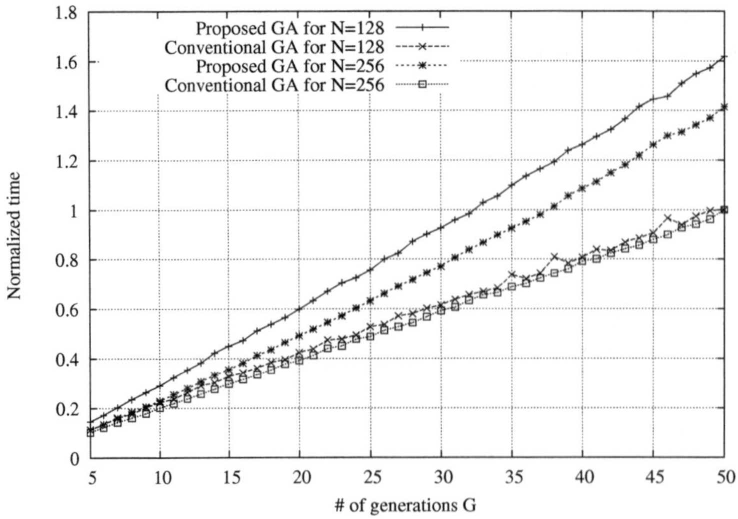

Fig. 6. Normalized calculation time versus the number of generations G

Table 2. p-values of Welch's t test for PAPR at $G = 50$

Method 1	Method 2	N	p-value
Prop. GA	GA with prop. mut.	128	6.0×10^{-4}
		256	4.3×10^{-3}
GA with prop. sel.	Conv. GA	128	2.2×10^{-3}
		256	4.4×10^{-11}

For fair comparisons, the proposed GA is compared with the conventional one when the computation time or the PAPR reduction performance is almost equivalent. Table 3 shows the PAPR reduction performance comparison when using approximately the same computation time. According to the result, for $N = 128$ and $N = 256$, the proposed GA is 0.11dB and 0.15dB better than the conventional one. Table 4 shows the computation time comparison when achieving approximately the same PAPR reduction performance. According to the result at IBO=3.5dB, for $N = 128$ and $N = 256$, the proposed GA is 30% and 34% faster than the conventional one, respectively.

5.2 Bit Error Rate Performance

The simulations assume an AWGN (Additive White Gaussian Noise) channel, where a white noise with SNR 20dB is added to the signal from the

Table 3. The PAPR reduction performance comparison when using approximately the same computation time

N	Method	G	Time	PAPR [dB]
128	Proposed GA	32	0.98	4.11
	Conventional GA	50	1.00	4.22
256	Proposed GA	37	0.98	4.61
	Conventional GA	50	1.00	4.76

Table 4. The computation time comparison when achieving approximately the same PAPR reduction performance

N	Method	G	Time	PAPR [dB]
128	Proposed GA	23	0.70	4.22
	Conventional GA	50	1.00	4.22
256	Proposed GA	26	0.66	4.76
	Conventional GA	50	1.00	4.76

amplifier. The used nonlinear power amplifier model has the following input-output characteristic.

$$F[\rho] = \frac{\rho}{(1+\rho^6)^{1/6}}, \tag{11}$$

where $F[\rho]$ is the gain of the amplifier and ρ is the ratio of the mean input amplitude to the saturation amplitude.

The proposed GA is compared with the conventional GA in terms of bit error rate (BER).

$$\text{BER} = \frac{\text{(Total number of error bits)}}{\text{(Total number of transmitted bits)}} \tag{12}$$

The result is shown in Fig.7, where IBO (Input Back-Off) defines the degree of nonlinearity of an amplifier and is given by the following equation.

$$\text{IBO} = \frac{\text{(Saturation amplitude)}}{\text{(Mean input power)}}. \tag{13}$$

A larger IBO degrades the energy efficiency of an amplifier.

For both $N = 128$ and $N = 256$, the proposed GA for any G provides a better BER than the conventional GA. When using approximately the same computation time as the conventional GA for $G = 50$, the proposed GA ($G = 32$ for $N = 128$ and $G = 26$ for $N = 256$) achieves approximately 59.0% and 41.9% better BER than the conventional GA. Their differences are statistically significant because p-values of Welch's t test are at most 2.9×10^{-7} and 1.1×10^{-6} for $N = 128$ and $N = 256$, respectively.

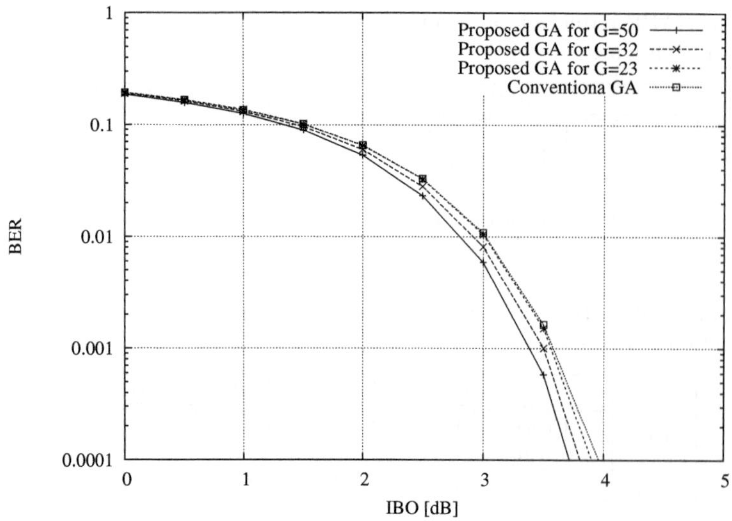

(a) For $N = 128$.

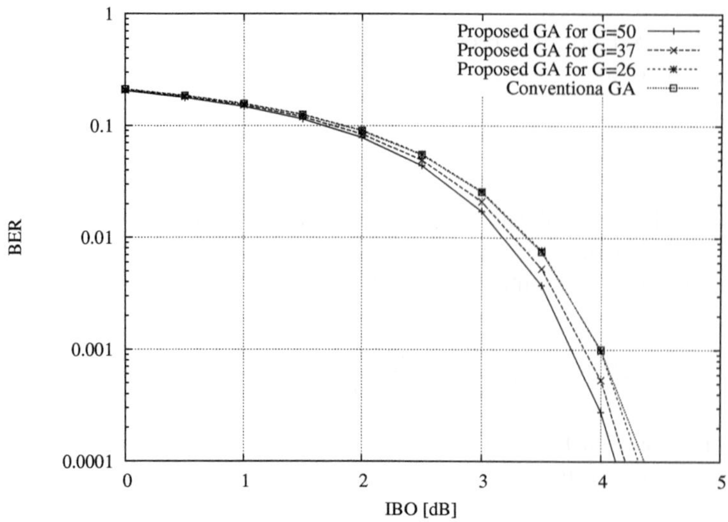

(b) For $N = 256$.

Fig. 7. BER versus IBO

6 Conclusions

In this paper, we proposed a selection scheme and a mutation operator and we applied them to GA for PAPR reduction of OFDM signal. The selection scheme is a hybrid of rank based selection and fitness proportionate selection, the mutation operator is based on the past record of generated chromosomes.

The simulation results showed that, compared with the conventional GA, the proposed GA can find a better solution in a shorter time. Therefore, the proposed GA can enhance the communication quality, the bit rate and energy efficiency in digital communication based on OFDM.

The proposed GA methods may be effective for other types of applications. Its investigation is one of our future works.

References

1. Weinstein, S., Ebert, P.: Data Transmission by Frequency-Division Multiplexing Using the Discrete Fourier Transform. IEEE Trans. on Communication Technology 19(5), 628–634 (1971)
2. Li, X., Cimini Jr., L.J.: Effects of Clipping and Filtering on the Performance of OFDM. IEEE Communications Letters 2(5), 131–133 (1998)
3. Armstrong, J.: New OFDM Peak-to-Average Power Reduction Scheme. In: Proc. of IEEE Vehicular Technology Conference, pp. 756–760 (2001)
4. Jones, A.E., Wilkinson, T.A., Barton, S.K.: Block Coding Scheme for Reduction of Peak to Mean Envelope Power Ratio of Multicarrier Transmission Scheme. Electronics Letters 30(25), 2098–2099 (1994)
5. Cimini Jr., L.J., Solenberger, N.R.: Peak-to-Average Power Ratio Reduction of an OFDM Signal Partial Transmit Sequences. IEEE Communications Letters 4(3), 86–88 (2000)
6. Tellado, J., Cioffi, J.: Peak Power Reduction for Multicarrier Transmission. In: Proc. of IEEE Communication Theory Mini-Conference, GLOBECOM 1998, pp. 219–224 (1998)
7. Ohta, M., Ueda, Y., Yamashita, K.: PAPR Reduction of OFDM Signal by Neural Networks without Side Information and its FPGA Implementation. IEEJ Trans. on Electronics, Information and Systems 126(11), 1296–1303 (2006) (in Japanese)
8. Shigei, N., Miyajima, H., Ozono, K.: Time-Efficient Genetic Algorithm for Peak Power Reduction of OFDM Signal. In: Lecture Notes in Engineering and Computer Science: Proceedings of The World Congress on Engineering and Computer Science 2010, WCECS 2010, San Francisco, USA, October 20-22, pp. 186–191 (2010)
9. Shigei, N., Miyajima, H., Ozono, K., Araki, K.: Acceleration of Genetic Algorithm for Peak Power Reduction of OFDM Signal. IAENG Int. Journal of Computer Science 38(1), 32–37 (2011)

Hands and Arms Motion Estimation of a Car Driver with Depth Image Sensor by Using Particle Filter

Norikazu Ikoma

Kyushu Institute of Technology, Japan
ikoma@ecs.kyutech.ac.jp

Abstract. Estimation of hands and arms motion in a car driving by utilizing a depth image sensor, specifically, KINECT of Microsoft Xbox 360, has been proposed. Compared with conventional researches using ordinary vision sensor, depth sensor provides rich information for the hands and arms in the scene. Especially, arms' regions detected by the depth sensor have been utilized to estimate the hands and arms motion more accurately than the conventional researches. As well as the increasing accuracy of the hands and arms region extraction, this paper proposes to incorporate some particles intentionally switching the left and the right of the hands in a framework of particle filter. This idea reduce the mistaken (opposite) determination of left and right and it will increase the opportunity to recover automatically from the opposite determination. Experiments over the recorded videos of vision and depth under a driving simulator environment show the efficiency of the proposed method.

1 Introduction

Safety driving support is one of the key technologies for near future development of traffic environment in order to become more sophisticated one. Not only outside of the car, but also behavior of a car driver is an important information to collect for achieving the safety driving support. As the car driver's behavior, head motion for observing the outside scene, hands motion for operating the car with respect to the direction of it, and foot motion for operating the car with respect to its acceleration are the typical motions to be captured as the car driver's behavior. In this paper, hands motion has been focused on as a target motion.

Particle filter [1] has been employed as an approach for capturing hands motion of a car driver due to its robustness by using many number of hypotheses, high accuracy by sequential Bayesian estimation, and real-time performance with the aid of parallel computation [3]. Conventional researches provide the ideas of tracking hands in captured video by using skin color ratio of designated square as a hand over the image [4], and arm direction estimation by evaluating the

S.-Y. Rhee et al. (eds.), *Soft Computing in Machine Learning*,
Advances in Intelligent Systems and Computing 273,
DOI: 10.1007/978-3-319-05533-6_8,

wrist region over the image [5]. Along with this line, this paper will try to improve the performance of hands motion estimation with the aid or depth image sensor.

Depth image sensor, specifically, KINECT of Microsoft Xbox 360, has been used for a sensor to capture the information for the motion estimation. A state of the art research [2] provides some results on capturing the hands motion that hands region over the image can easily be extracted by using the depth information obtained from the sensor. In this paper, not only extracting the hands region, but also to extract the region of arms by utilizing the depth information. Also, extracted arms region is used to determine left or right of the hand/arm by introducing a likelihood evaluation for the arm region. Additionally to the increasing accuracy of the hands and arms region extraction, this paper proposes to incorporate some particles intentionally switching the left and the right of the hands in a framework of particle filter. This idea reduce the mistaken (opposite) determination of left and right and it will increase the opportunity to recover automatically from the opposite determination.

2 Depth Sensor in Driving Simulator

Fig.1 shows a setup of the depth image sensor in driving simulator environment above the car driver's head. KINECT of Microsoft Xbox 360 has been used as a depth image sensor, with the aid of OpenNI library for developing the natural user interface. Videos of ordinary camera image and depth image are captured by a program developed in C language with OpenNI, as well as OpenCV library for developing image processing, displaying the images, and saving the recorded videos in a motion video files.

Extraction of hand/arm region over the image by combining the visual image and depth image has been conducted for each image frame in the frame rate around 30 [fps:frame per second]. The extraction is based on color information

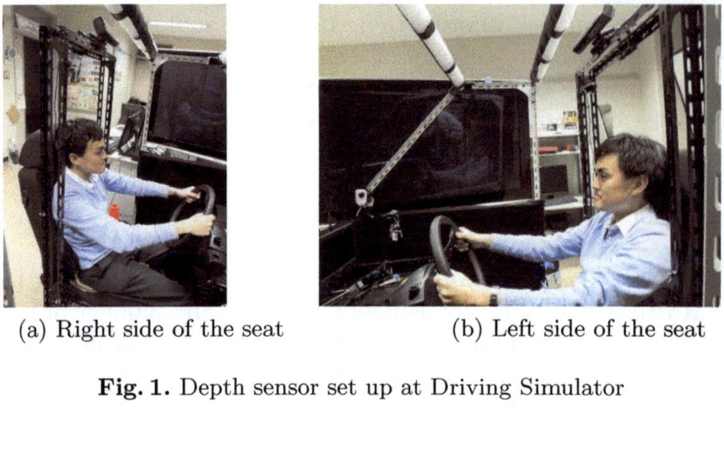

(a) Right side of the seat (b) Left side of the seat

Fig. 1. Depth sensor set up at Driving Simulator

and depth information for each pixel. For far or equal distance pixel than the distance to the steering, it is extracted as hand/arm region when its color is within skin color interval. Where YCrCb color space is used to determine the skin color. The nearer distance pixel than the distance to the steering is extracted as hand/arm region without consideration of its color. Then a binary image consisting of extracted hand/arm region having pixel value of 255 in gray scale, 0 for the other region has been obtained.

3 Model

A state space model of hands motion has been explained here for capturing the hands motion of a car driver within a framework of state estimation via particle filter. A state vector at discrete time k consists of left and right hands components such as

$$\mathbf{x}_k = \left[\mathbf{x'}_k^L, \ \mathbf{x'}_k^R\right]' \tag{1}$$

where each component contains grasping position of the steering in angle, $\theta_k^\bullet$, and angular direction to the wrist region from the hand region, $\varphi_k^\bullet$, for $\bullet \in \{L, R\}$, such that

$$\mathbf{x}_k^L = \left[\theta_k^L, \ \varphi_k^L\right]', \ \mathbf{x}_k^R = \left[\theta_k^R, \ \varphi_k^R\right]'. \tag{2}$$

Time evolution of the state is represented in a state equation of a random walk with respect to the state vector such that

$$\mathbf{x}_k = \mathbf{x}_{k-1} + \mathbf{v}_k \ \ , \ \ \mathbf{v}_k \sim N(\mathbf{0}, \mathbf{Q}). \tag{3}$$

The system equation in eq.(3) is re-written in a probabilistic form, which is called "system model"

$$f(\mathbf{x}_k|\mathbf{x}_{k-1}) = N(\mathbf{x}_k; \mathbf{x}_{k-1}, \mathbf{Q}), \tag{4}$$

where $N(\mathbf{x}; \mathbf{m}, \mathbf{\Sigma})$ denotes probability density function of normal distribution with mean vector $\mathbf{m}$ and covariance matrix $\mathbf{\Sigma}$).

One another element of a state space model is "observation model", which is used to evaluate likelihood value in a particle filter framework with respect to a given sensor information at time k, denoted by $\mathbf{I}_k$. Note that $\mathbf{I}_k$ is the binary image consisting of extracted hand/arm region with 255 pixel value and 0 for the other region as explained in section 2.

Observation model for hands motion estimation consists of likelihoods for left and right hands such that

$$h(\mathbf{I}_k|\mathbf{x}_k) = h_L(\tilde{\mathbf{I}}_k^L|\mathbf{x}_k^L)h_R(\tilde{\mathbf{I}}_k^R|\mathbf{x}_k^R), \tag{5}$$

where $\tilde{\mathbf{I}}_k^L$ and $\tilde{\mathbf{I}}_k^R$ are sensor information for left and right hands respectively by eliminating the other hand/arm region based on the mean value of one-step-ahead prediction of the state vector using the system equation in eq.(3).

Each hand likelihood has a multiplicative combination of hand region's likelihood, wrist region one, and arm region one, such that

$$h_\bullet(\tilde{\mathbf{I}}_k^\bullet|\mathbf{x}_k^\bullet) = h_\bullet^{hand}(\tilde{\mathbf{I}}_k^\bullet|\mathbf{x}_k^\bullet)h_\bullet^{wrist}(\tilde{\mathbf{I}}_k^\bullet|\mathbf{x}_k^\bullet)h_\bullet^{arm}(\tilde{\mathbf{I}}_k^\bullet|\mathbf{x}_k^\bullet). \tag{6}$$

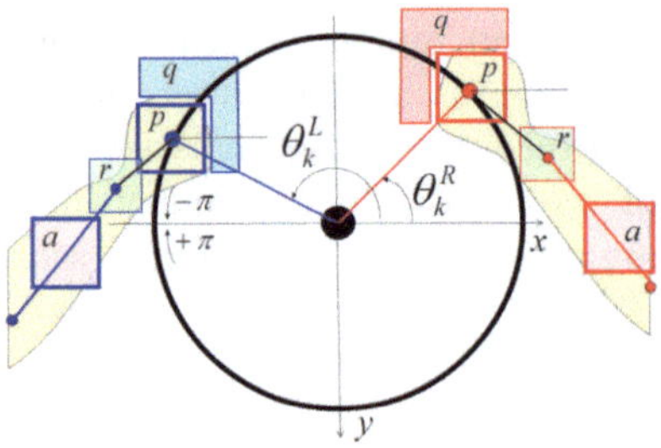

(a) Likelihood for both hand

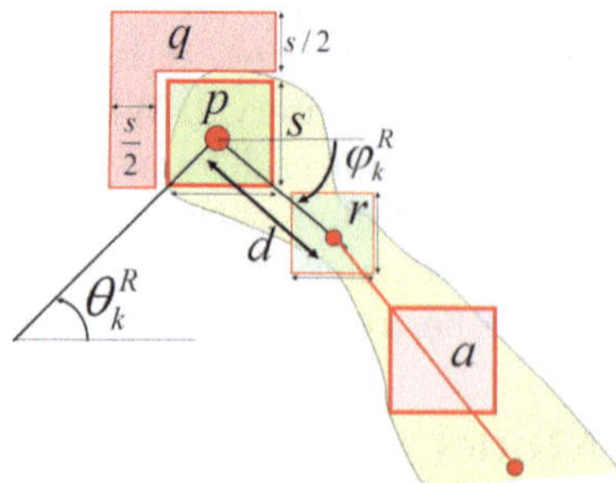

(b) Detailed likelihood for right hand

Fig. 2. Likelihood computation for hand, wrist, and arm

Likelihood of these regions are depicted in Fig.2. Hand region's likelihood consists of combination of positive use of palm region likelihood p and negative use of top of the palm region likelihood q such that

$$h_{\bullet}^{hand}(\tilde{\mathbf{I}}_k^{\bullet}|\mathbf{x}_k^{\bullet}) = p(\tilde{\mathbf{I}}_k^{\bullet}, \mathbf{x}_k^{\bullet}) \times \left\{1 - q(\tilde{\mathbf{I}}_k^{\bullet}, \mathbf{x}_k^{\bullet})\right\}. \tag{7}$$

Where, palm region is a square of size S with its position being determined by $\theta_k^{\bullet}$, and top of the palm region is determined as shown in Fig.2. Likelihood of each region is evaluated by counting the number of pixels corresponding to the extracted hand/arm region and its ratio to the total number of the pixels in the region is employed as the likelihood value. Such evaluation of likelihood will be used for the regions of wrist and arm in the followings.

Wrist region's likelihood consists of positive use of wrist region likelihood r as

$$h_{\bullet}^{wrist}(\tilde{\mathbf{I}}_k^{\bullet}|\mathbf{x}_k^{\bullet}) = r(\tilde{\mathbf{I}}_k^{\bullet}, \mathbf{x}_k^{\bullet}), \tag{8}$$

where the wrist region is determined by $\varphi_k^{\bullet}$ with fixed distance from a center of the palm region as shown in Fig.2. The wrist region is square with size smaller than that of palm region with the ratio, say 0.9.

Arm region's likelihood consists of positive use of arm region likelihood a as

$$h_{\bullet}^{arm}(\tilde{\mathbf{I}}_k^{\bullet}|\mathbf{x}_k^{\bullet}) = a(\tilde{\mathbf{I}}_k^{\bullet}, \mathbf{x}_k^{\bullet}). \tag{9}$$

The arm region is square with size the same as the palm region. Position of the arm region is determined as the central point between the center of wrist region and a fixed point over the image that corresponds to the root of the arm in the image.

4 Estimation

Particle filter [1] has been used for the state estimation. It uses many number, M, of realizations in the state space to approximate the posterior distribution of the state for given sequence of the sensor signals up to the current time

$$\left\{ \mathbf{x}_k^{(i)} \right\}_{i=1}^{M} \approx p(\mathbf{x}_k | \mathbf{I}_{1:k}), \tag{10}$$

where $\mathbf{I}_{1:k}$ denotes collection of sensor information from initial time 1 to current time k.

To update the set of particles from previous time $k-1$ to current time k, following the three steps called "Sampling", "Importance weight", and "Re-sampling" are conducted in this order. First, "Sampling" is conducted by drawing a particle of current time step using system model with a particle of previous time step in the conditional part

$$\mathbf{x}_k^{(i)} \sim f(\mathbf{x}_k | \mathbf{x}_{k-1}^{(i)}) \ . \ i = 1, 2, \cdots, M. \tag{11}$$

Then, "Importance weight" is evaluated for the draw by observation model

$$w_k^{(i)} = h(\mathbf{I}_k | \mathbf{x}_k^{(i)}). \ . \ i = 1, 2, \cdots, M. \tag{12}$$

Finally, "Re-sampling" is conducted by proceeding a draw with replacement with respect to the drawn particles in eq.(11) with probability proportional to the weight. Refer [1] for more details.

Some elaborations have been employed in the estimation procedure to discriminate left and right arms and determination of existence of each hand. Refer [5] for details. Additionally to the elaborations, this paper proposes to incorporate some particles intentionally switching the left and the right in the state vector in eq.(1). This idea reduce the mistaken (opposite) determination of left and right and it will increase the opportunity to recover automatically from the opposite determination.

5 Experiments

For recorded videos of visual image and depth image, the proposed method and conventional methods based on visual image only [5] and depth image [2] have been examined. For two test subject A and B, three runs of an elaborated road in the driving simulator have been conducted and videos have been recorded. Especially, for the 1st run for each test subject, there are several mistaken (opposite) determinations of left and right, so these scenes are focused on here. Some resulting frames for these scenes have been shown in Fig.3 for test subject A, and Fig.4 for test subject B. Each frame of the figures, shows extracted hand/arm region in binary image at the left panel, hands motion estimation result in mean of the posterior distribution in the middle panel, and all particles in the right panel.

Through human observations by the author, less mistaken (opposite) determinations of left and right than the conventional method has been confirmed. Furthermore, the proposed method automatically recovered from such mistaken determinations within several frames.

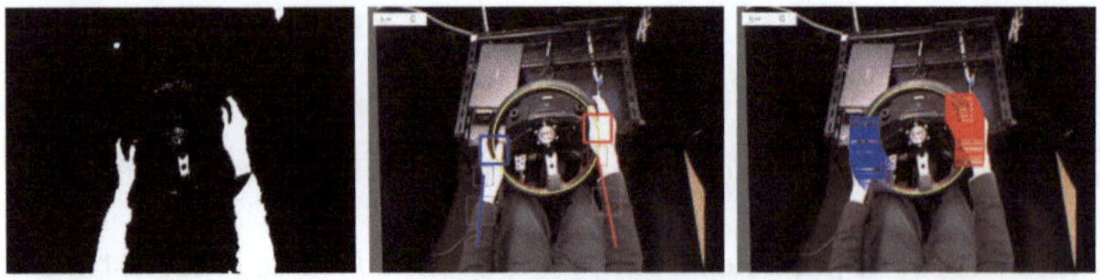

(a) 0-th frame

(b) 50-th frame

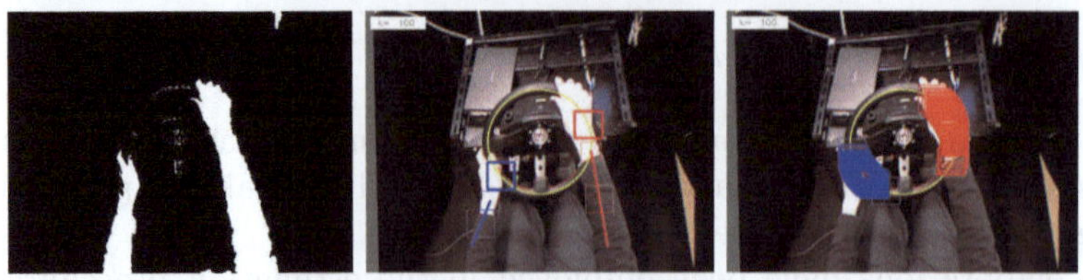

(c) 100-th frame

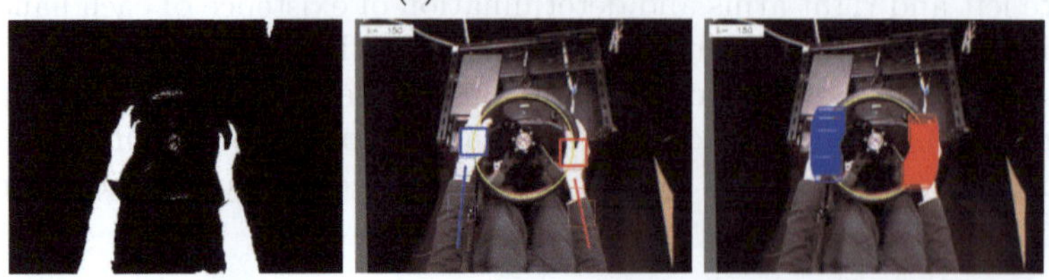

(d) 150-th frame

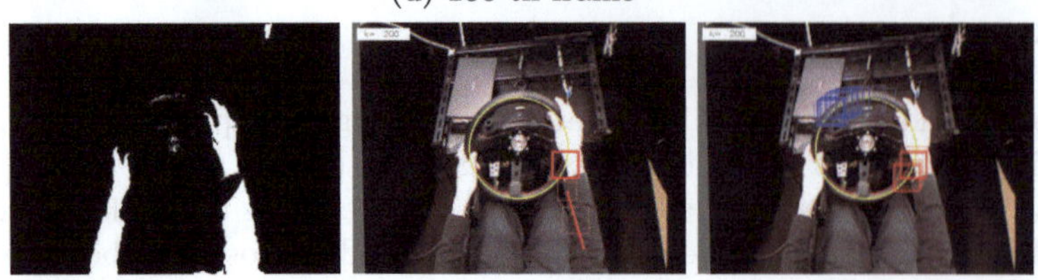

(e) 200-th frame

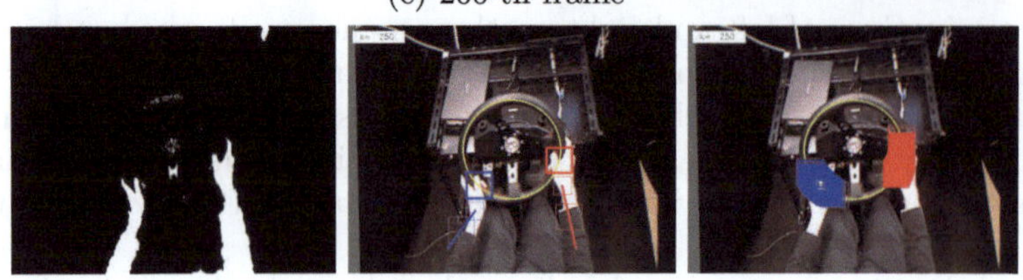

(f) 250-th frame

Fig. 3. Result : test subject A, 1st run

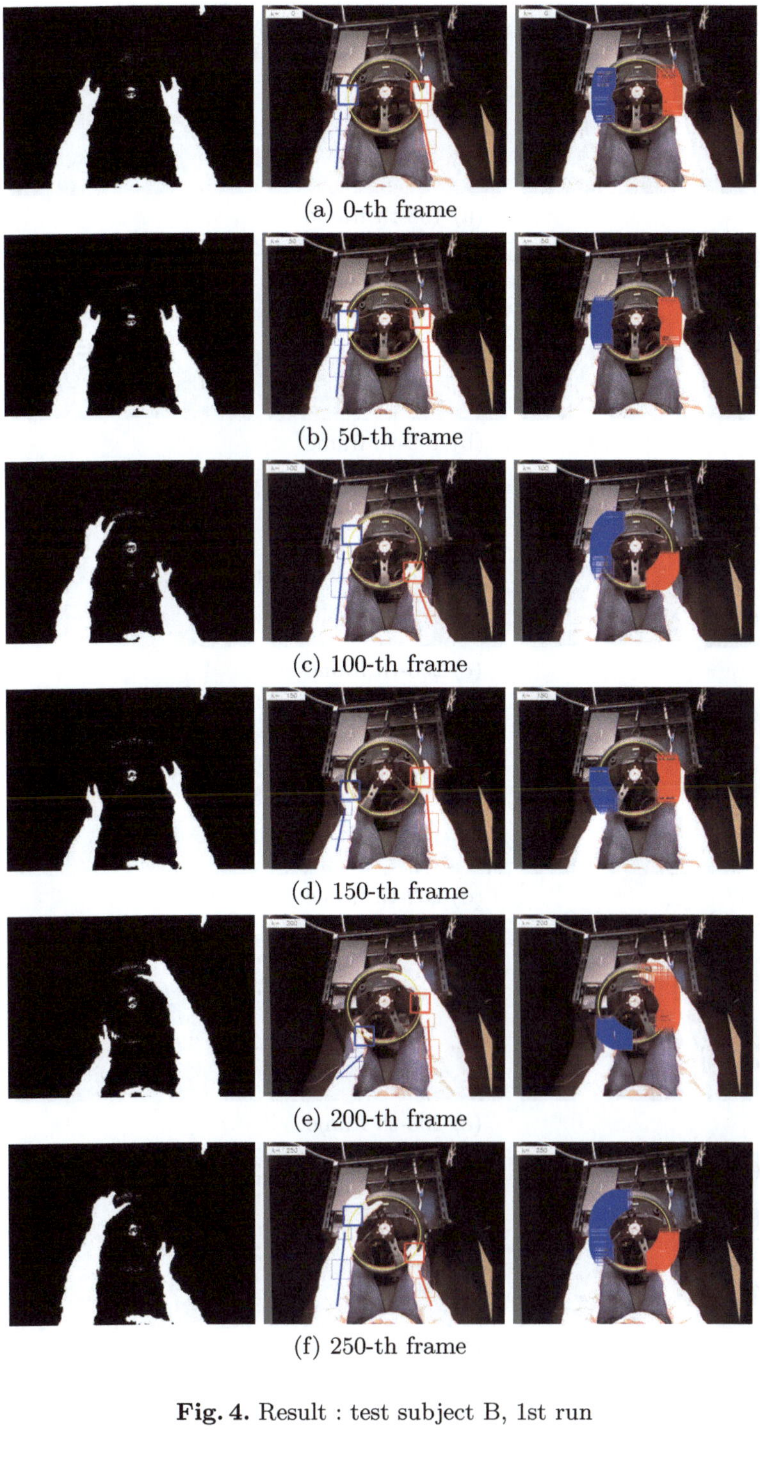

(a) 0-th frame
(b) 50-th frame
(c) 100-th frame
(d) 150-th frame
(e) 200-th frame
(f) 250-th frame

Fig. 4. Result : test subject B, 1st run

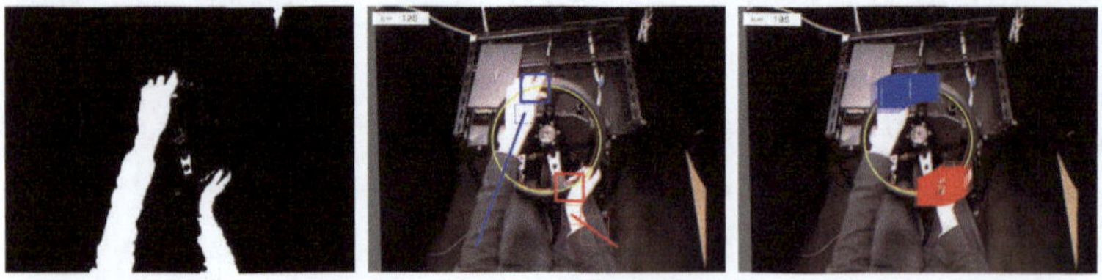

(a) 198-th frame

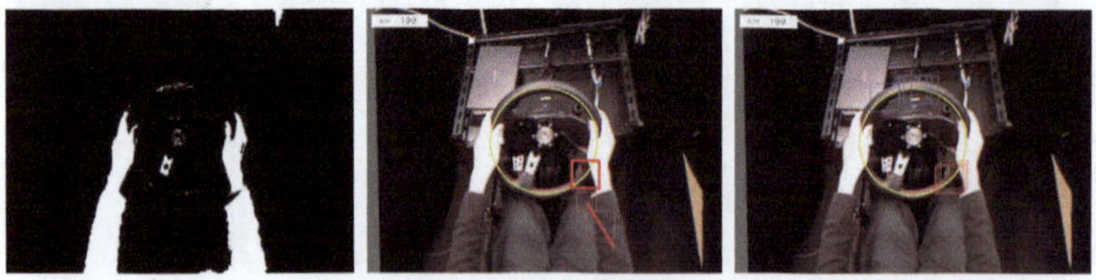

(b) 199-th frame

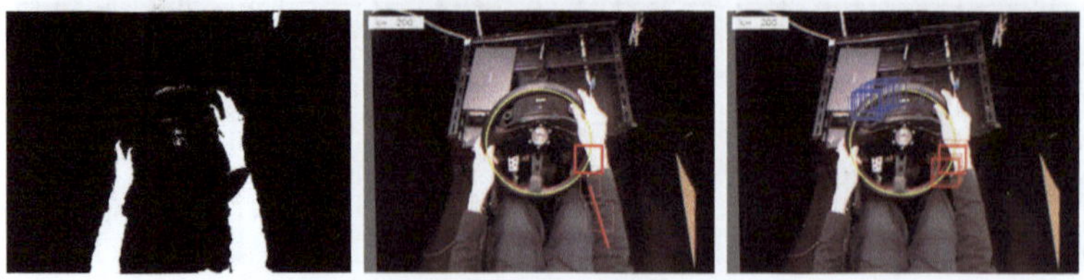

(c) 200-th frame

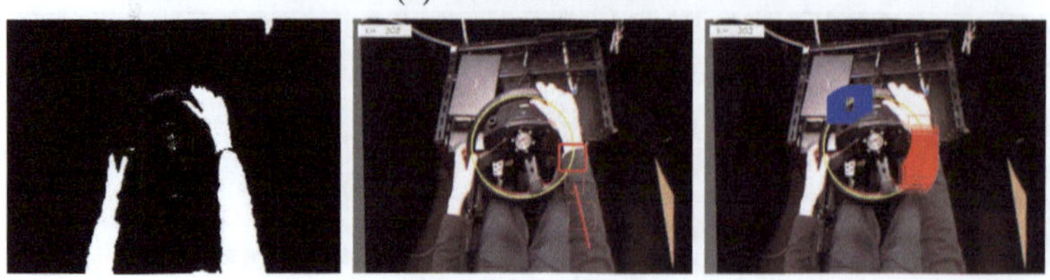

(d) 202-th frame

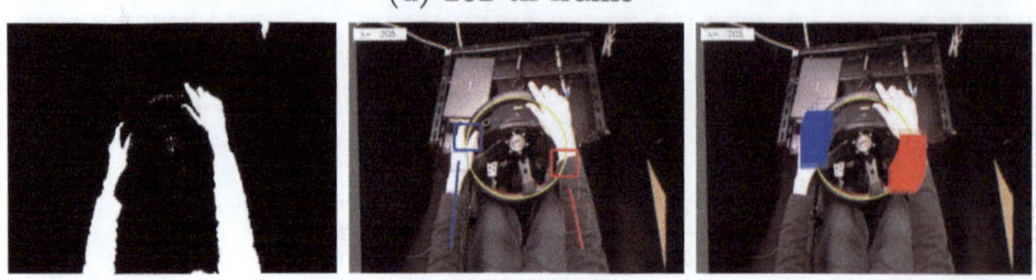

(e) 205-th frame

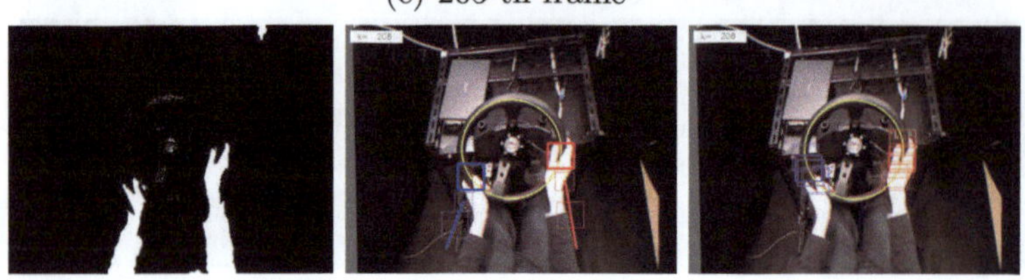

(f) 208-th frame

Fig. 5. Detailed result : test subject A, 1st run, 198-208 frames

In fact, (e) 200-th frame of Fig.3 has a lost of left hand due to fast motion of hands (or lack of image frames due to delay of processing in PC), however it can recover from the lost automatically as detailed frames are shown in Fig.5. The figure shows several important frames from 198-th frame to 208-th frame. At the 199-th frame, fast motion of hands (or lack of real-time capture of image frame due to delay of processing in PC) occurred, then it leads to a lost of left hand. However, it can recover to the correct determination of left and right with proper tracking of the hands as time goes by. This is a typical robust feature of particle filter, which is one of the probabilistic methods, that utilises many number of candidates (particles) for tracking. On the other hand, deterministic methods may fail for such cases due to their single candidate feature.

6 Conclusion

Estimation of hands and arms motion in a car driving by utilizing a depth image sensor, specifically, KINECT of Microsoft Xbox 360, has been proposed. Increasing accuracy of extraction of hands/arms region due to the use of depth information as well as an elaborated particle filter procedure to incorporate some particles intentionally switching the left and the right of the hands, the proposed method achieves better performance of hands motion estimation without having many mistaken (opposite) determinations of left and right as in conventional methods through experiments over the recorded videos of vision and depth under a driving simulator environment.

Future works include parallel implementation in GPGPU for fast and more accurate estimation with more number of particles, as well as real-time implementation of the proposed method. Real car environment is also interesting and important issue for more realistic evaluation of the methods.

Acknowledgment. This work is partially supported by ISM cooperative research program (2013-ISM-CRP-2015).

References

1. Doucet, A., de Freitas, N., Gordon, N.J. (eds.): Sequential Monte Carlo Methods in Practice. Springer, New York (2001)
2. Ikoma, N.: Tracking of Car Driver's Hands in Depth Image Sensor by Particle Filter. In: 8th International Conference on Innovative Computing, Information and Control (ICICIC 2013) (2013) (to appear)
3. Ikoma, N., Ito, T.: GPGPU implementation of visual tracking by particle filter with pixel ratio likelihood. In: 2012 IEEE/SICE International Symposium on System Integration (SII), pp. 889–894 (2012)
4. Ikoma, N.: Visual tracking of both hands of car driver by particle filter. In: 5th Int'l Conf. on Soft Computing and Intelligent Systems and 11th Int'l Sympo. on advanced Intelligent Systems (SCIS & ISIS 2010), pp. 1547–1552 (2010)

5. Ikoma, N.: Real-Time Motion Estimation of Car Driver's Hands and Arm's Direction in Vision under Possible Mutual Occlusion by Particle Filter. In: 6th Int'l Conf. on Soft Computing and Intelligent Systems and 13th Int'l Sympo. on advanced Intelligent Systems (SCIS & ISIS 2012), pp. 701–704 (2012)

SNR Improvement of an Optical Wave Microphone Using a Wavelet Network

Takashi Samatsu and Yoshito Sonoda

Graduate School of Industrial Engineering,
Tokai University
9-1-1 Toroku, Kumamoto 862-8652, Japan
samatsu@tokai.ac.jp,
ysonoda@ktmail.tokai-u.jp

Abstract. In this paper, we discuss an approach of S/N ratio improvement for an optical wave microphone. With the optical wave microphone, ultra-weak diffraction light; which is resulting from phase modulation from a sound wave, is converted to an electrical signal using a light detector. In this way the optical wave microphone can detect the sound without making any contact with others. Signal intensity could be increased or S/N ratio could be improved with (1) optical method, (2) method using electrical circuit, or (3) method using information processing. In this study, we adopted wavelet network as one of the methods using information processing. As a result we succeeded noise reduction at low frequency using only proposed algorithm, not by using combination of various filters.

Keywords: optical wave microphone, laser, wavelet network, convex wavelet.

1 Introduction

Optical wave microphone is a method of detecting sound directly from light, and has many advantages of being resistant to high frequency measurement without disturbing the sound field etc., compared with the conventional vibrating membrane microphone [1-6]. However verification of sound detection theory using interaction between light and sound, and clarification of the basic characteristics have been mainly focused so far, and S/N ratio and the measurement sensitivity, important factors for practical application, have not been fully concerned. One of the major challenges for the practical application is to increase the signal intensity and to improve the S/N ratio. We especially experiments how to improve the microphone's sensitivity and the S/N ratio, and those possibilities. Wavelet network is adopted to the signal processing part as one of the ways to improve S/N ratio. This network is fused with functional link net and wavelet theory [7, 8], and it has high non-linear description capability [9, 10]. In addition, using a simple wavelet to a base function can acquire high generalization ability. This ability extracts original signal component, which leads to S/N ratio improvement of the optical wave microphone.

S.-Y. Rhee et al. (eds.), *Soft Computing in Machine Learning*,
Advances in Intelligent Systems and Computing 273,
DOI: 10.1007/978-3-319-05533-6_9, © Springer International Publishing Switzerland 2014

2 Optical Wave Microphone

2.1 Principle and Theory

The fundamental principle and theory of the optical wave microphone have been already established [1], but an abstract is shortly introduced in this chapter to explain the background of research and to use the theoretical equation in the later discussion for experimental results.

Fig. 1 shows abstract of the model for theoretical analysis. When an incidence probing laser beam crosses a sound wave, diffraction light waves are generated and propagate with and in the penetrating beam through the Fourier optical system and reach the detection plane, which is set in the back focal plane of a receiving lens. The diffracted light is homodyne-likely detected there by using the penetrating laser light as a local oscillating power. In such condition, the intensity of diffracted light of higher orders and multiple diffractions are much smaller than that of the first order diffraction light and can be neglected. In the practical machine, the electrical output from a photo detector includes both the DC component and the AC component relating to diffraction light Doppler-shifted by sounds. The DC component is removed in the electric circuit and only the 1st AC component is used. The spatial intensity of diffraction light signal for the theoretical model shown in Fig. 1 is given by the following equation [2, 3].

$$I_{ac} = I_0 \Delta\varphi_0 \exp(-u^2)\left[\exp\{-(u-\theta)^2\} - \exp\{-(u+\theta)^2\}\right]\cos\omega_a t \tag{1}$$

where $I_0 = (2P_0/\pi w_f^2)\exp\left[-2(y_f/w_f)^2\right]$ [W/m^2], $\Delta\varphi = k_i(\mu_0 - 1)\Delta Z \Delta p/\gamma p$, μ_0: refractive index of air, γ: specific heat ratio, ΔZ: width of sound, p: atmospheric pressure, Δp: sound pressure, k_i: wave number of laser light, ω_a: angular frequency of sound wave, P_0: laser power, $u=x_f/w_f$: the normalized x-coordinate in the back focal plane, $\theta=k_a w_0/2$: the normalized wave number, k_a: wave number of sound wave, w_0: radius of laser beam waist in sound incident region, w_f, x_f, y_f: radius of the beam cross section, x-coordinate and y-coordinate in the observing plane, respectively.

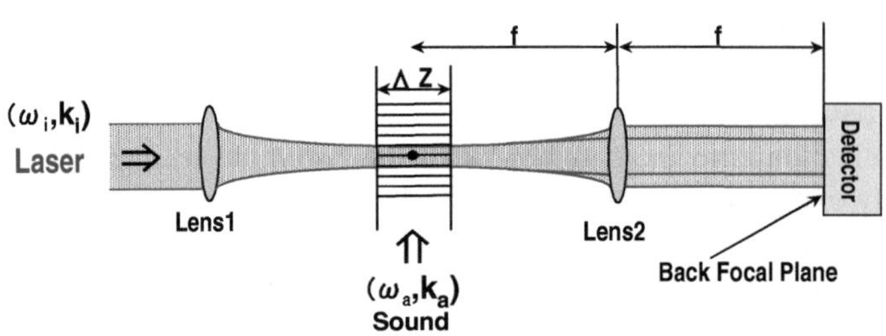

Fig. 1. Optical setup for theoretical analysis

Based on the equation, numerical calculations of the diffraction pattern are carried out, in which a visible laser was assumed as a probing laser beam. Fig. 2 shows examples of spatial distributions of the intensity and the phase of the diffraction light pattern. The spatial profile of diffracted light pattern (I) oscillating at the sound frequency has two peaks, which spatial positions do not change with frequency in the audio-wave or the low frequency ultrasonic band. On the other hand, the temporal phase difference (Φ) between the right and left diffraction patterns oscillating at ω_a is π, as shown in Fig. 2(right).

The equation (1) under the condition of $\theta << 1$ (Laser beam diameter w_0 is small and sound wave length λp is short) is approximately represented by following equation:

$$I_{ac} \approx 4I_0 \Delta\varphi_0 e^{-2u^2} u\theta \cos\omega_a t \tag{2}$$

where $\theta = k_p w_0 / 2 = \pi w_0 f_p / c_p$, f_p: sound wave frequency, c_p: speed of sound. The optical signal intensity I is theoretically proportional to the frequency of sound wave. In application to sound measurement, the frequency response from the optical wave microphone system is made flat over the whole frequency band by an electric signal processing circuit.

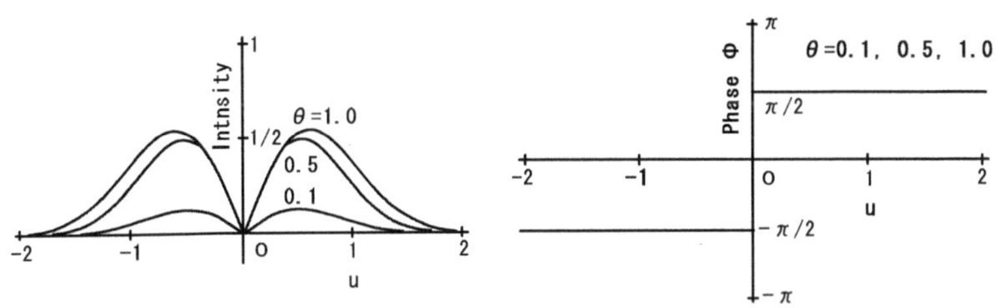

Fig. 2. Theoretical profiles of diffracted light distribution. A spatial intensity profile(left) and a Spatial phase profile(right).

2.2 Property and Merits

Based on the above principle, we can construct various types of optical wave microphones, which can directly detect acoustic waves and have many advantages as follows: (1) Unlike the conventional microphone, sound measurement is applicable under conditions of high electric and/or magnetic fields, and even under the condition of airtight chamber, if a small optical window exists. (2) Any lasers from visible to near infrared wavelengths can be used. Either visible or invisible laser microphone can be constructed depending on the users' needs. (3) The sound receiving part or laser beam of the optical wave microphone can be constructed in one to three dimensions, by which we can offer various directivities and sizes from a compact type to an ultra-large antenna type called as the macrophone. (4) The sound signal could be optically amplified and transmitted by an optical fiber without disturbance of electromagnetic noise.

3 Wavelet Network

3.1 Structure of Wavelet Network

In this section, several types of wavelet networks are introduced. Fig. 3 shows a simple wavelet network, where the input signal X_k is supplied (as commonly) to each basis function. The output of this network is represented by a weighted sum of each basis function:

$$f_k(X_k) = \sum_{i=1}^{n} W_i \psi_i(X_k). \tag{3}$$

This network represents a single-input single-output nonlinear system.

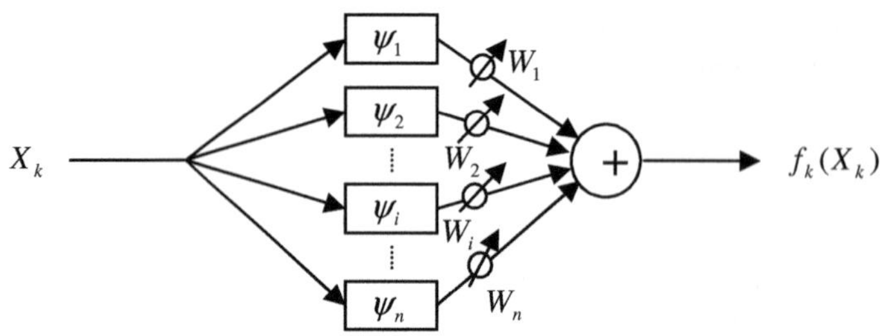

Fig. 3. A simple wavelet network

An input-noncorrelated wavelet network consists of the combination of the simple wavelet networks of Fig. 3. The output Y is represented by:

$$Y = \sum_{k=1}^{s} f_k(X_k) = \sum_{k=1}^{s}\sum_{i=1}^{n} W_i^k \psi_i(X_k). \tag{4}$$

The relation between the input X_k $(k = 1, 2, \cdots, s)$ and the output Y is embedded in the functions f_k $(k = 1, 2, \cdots, s)$ by learning. The output Y is dependent on all the weights W_i^k. Therefore, if we use the squared error criterion for the settlement of weights, the local minimum, as in the case of neural networks, does not exist. Indeed, the error criterion is based on a quadratic form with respect to the weights, and thus the optimal weights can be found.

Steepest descend method is adopted to renew the connection weights in the network [9, 10].

3.2 Learning Algorithms

In the traditional steepest descend method, the learning rate is a constant value and is determined by a trial and error procedure. This may cause a slow learning, an oscillatory learning, or a divergent learning. To solve this problem, we present a modified steepest descend method. The method is then used to train the wavelet network. The network weights W_i of the wavelet network are determined by the modified method.

The renewal of weights in the traditional steepest descend method is given by the equation:

$$\Delta W_i = -G \frac{\partial E}{\partial W_i} \tag{5}$$

with

$$E = \frac{1}{2} \cdot \sum_{j=1}^{m} \left(f_j - \hat{f}_j \right)^2 , \tag{6}$$

Here, G denotes the learning rate, f_j is an actual output of the network, and $\hat{f}_j$ is its training data. The renewal of weights is modified as:

$$\Delta W_i = -G(l) \cdot \mathrm{sgn}\left(\frac{\partial E}{\partial W_i} \right), \tag{7}$$

where

$$\mathrm{sgn}(x) = \begin{cases} 1 & (x \geq 0) \\ -1 & (x < 0). \end{cases} \tag{8}$$

In this learning procedure, the learning rate $G(l)$ $(l = 1, 2, \cdots)$ is variable, and it is initialized for each weight, in each iteration. The initial value $G(0)$ is assigned to be unity. A tentative value $G'(l+1)$, that is assigned to be equal to $G(l)$, is adopted first to calculate the weight $W_i'(l+1)$ and the error $E'(l+1)$ tentatively. If this tentative error $E'(l+1)$ is less than the error $E(l)$ at the previous step, the value $G(l+1)$ is not changed; otherwise, it is assigned to be half of the previous value $G(l)$:

$$G(l+1) = \begin{cases} G(l) & (E'(l+1) < E(l)) \\ \frac{1}{2} G(l) & (\mathit{otherwise}). \end{cases} \tag{9}$$

This iteration is repeated until the condition $|E(l+1)-E(l)| \le \delta$ is satisfied, where δ is given in advance.

3.3 Convex wavelet and Its Assignment

For a mother wavelet $\psi(t)$, a convex wavelet, having high generalization capability ([8], [9]) is adopted:

$$\psi(t) = \begin{cases} \cos \pi t & (-0.5 \le t \le 0.5) \\ 0 & (otherwise). \end{cases} \tag{10}$$

This convex wavelet is compactly supported, and thus the learning speed is increased. Furthermore, because this convex wavelet has only one extreme value over the support, a signal reconstructed with these wavelets is not oscillatory. The generalization capability is improved by employing this type of convex wavelet [9, 10].

The mother wavelet $\psi(t)$ generates basis functions:

$$\psi_{a,b}(t) = \psi(at - b), \tag{11}$$

where a (= 0, 1, …, M) is a scaling parameter and b (= 0, 1, …, a) is a shifting parameter.

These basis wavelets are assigned over the normalized fluctuation range. The numbers of wavelet bases is $1, 2, \cdots, M+1$ for the scaling parameter $a = 0, 1, \cdots, M$, respectively. The proof of convergence of this wavelet series is given in [10].

4 Non-linear Filter Using a Wavelet Network

4.1 Experimental Method

Fig. 4 shows a structure of wavelet network for non-linear filter. If time-series signal is input to an input-noncorrelated wavelet network, output from the network $\hat{X}_{p+1}$ is given as follows:

$$\hat{X}_{p+1} = \sum_{k=1}^{s} f_k(X_{p-k+1}) = \sum_{k=1}^{s}\sum_{a=0}^{M}\sum_{b=0}^{a} W_{a,b}^{k}\psi_{a,b}(X_{p-k+1}), \tag{12}$$

where $W_{a,b}^{k}$ is a connection weight. After learning, the relation between $X_p, X_{p-1}, \cdots, X_{p-s+1}$ and X_{p+1} is embedded into $f_1, f_2, \cdots, f_s$ as the set of connection weights. Learning modifies a connection weight $W_{a,b}^{k}$ of each f_k using proposed steepest descend method to minimize the error function with X_{p+1} and the output from the network $\hat{X}_{p+1}$. Repeating this renewing process allows the connection weight to converge to a most appropriate value. After learning the connection weight is fixed and signals are output from the network sequentially.

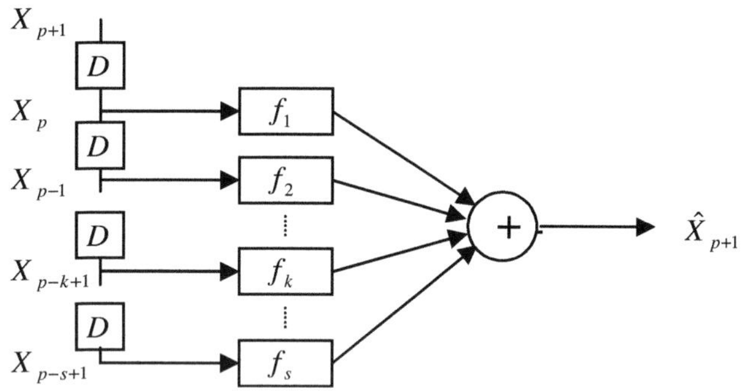

Fig. 4. A wavelet network for a non-linear filtering

4.2 Experimental Result

To confirm the effect of noise reduction, the sound (pure 10kHz tone, sound pressure level: 90dB) was measured with the optical wave microphone. Fig.5 shows its frequency spectrum. A band filter, which was adopted in advance to reduce oscillation noise and high-frequency noise, affected the area below 100Hz and above 20kHz. Measured signal is input to the wavelet network and filtering effect was confirmed. Fig.6 shows the frequency spectrum of output from the network, with network order s=5 and maximum value of scaling parameter M=8. A calculation time is around 4 minutes (CPU: 1.7GHz Intel Core i7, Memory: 8GB). Compared with Fig.5, noise could be reduced by around 2-3dB near the test signal, and by around 13-25dB at 100-2kHz. Effectiveness of this method was proved.

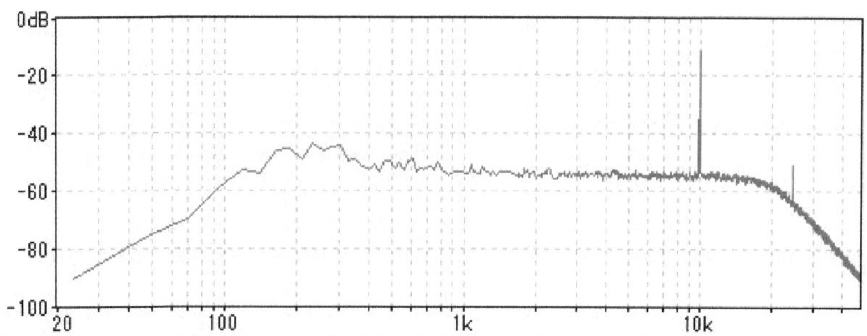

Fig. 5. Frequency spectrum observed by an optical wave microphone (Test signal: pure 10kHz tone, 90dB)

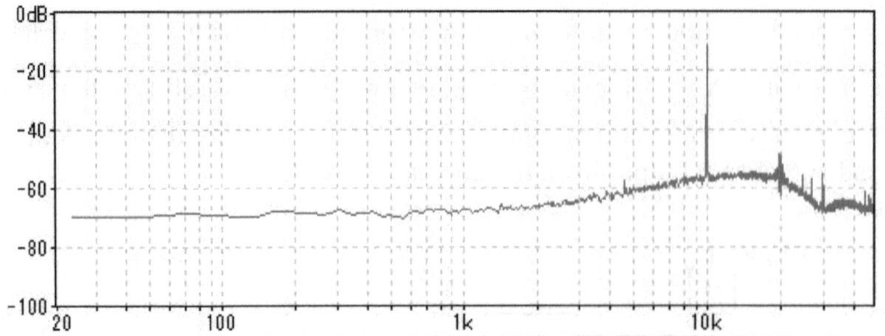

Fig. 6. Frequency spectrum of the output of a wavelet network

5 Conclusion

Wavelet network with convex wavelet was adopted to signal processing part to improve S/N ratio of optical wave microphone. Noise could be reduced by around 2-3dB near the input sound (pure 10kHz tone, 90dB), and by around 13-25dB at 100-2kHz. S/N ratio improvement, specially at low frequency, is the most challenging matter of optical wave microphone, therefore the proposed method could be one of the effective methods. In near future, we try to put this method to practical use as all-around noise reduction method with the proposed algorithm only, not by using various filters. We also try to establish most appropriate setting of the wavelet network by evaluating and verifying signals input at various sound pressures or frequencies. Performance comparison with other filters will also be conducted.

Acknowledgements. The authors acknowledge the supports of the Grant–in-Aid for Scientific Research (C; No.22560429) from Japan Society for the Promotion of Science.

References

1. Sonoda, Y., Akazaki, M.: Measurement of Low-Frequency Ultrasonic Waves by Fraunhofer Diffraction. Jpn. J. Appl. Phys. 33(pt. 1, 5B), 3110–3114 (1994)
2. Sonoda, Y.: Direct detection of acoustic waves by laser light diffraction and proposals of the optophone. In: Proc.16th Int. Cong. on Acoust. and 135th Meet. of Acoust. Soc. America, vol. 1, pp. 427–428 (1996)
3. Goodman, J.W.: Introduction to Fourier Optics, 3rd edn. Roberts & Company Publishers (2005)
4. Evans, D.E., von Hellermann, M., Holzhauer, E.: Fourier optics approach to far forward scattering and related refractive index phenomena in laboratory plasmas. Plasma Phys. 24, 819–834 (1982)
5. Sonoda, Y., Suetsugu, Y., Muraoka, K., Akazaki, M.: Applications of the Fraunhofer-diffraction method for plasma-wave measurements. Plasma Phys. 25, 1113–1132 (1983)

6. Tanaka, K., Nogami, G., Sonoda, Y.: Measurement of Audible Acoustic Wave by Laser Light. The Transactions of the Institute of Electrical Engineers of Japan 122-E(7), 362–368 (2002)
7. Morlet, J., Arens, G., Fourgeau, E., Giard, D.: Wave Propagation and Sampling Theory - Part 1: Complex Signal and Scattering in Multilayered Media. Geophysics 47(2), 203–221 (1982)
8. Morlet, J., Arens, G., Fourgeau, E., Giard, D.: Wave Propagation and Sampling Theory - Part 2: Sampling Theory and Complex Waves. Geophysics 47(2), 222–236 (1982)
9. Yamakawa, T., Uchino, E., Samatsu, T.: Wavelet Neural Networks Employing Over Complete Number of Compactly Supported Non-Orthogonal Wavelet and Their Applications. In: Proc. IEEE Int. Conf. on Neural Networks, pp. 1391–1396 (1994)
10. Yamakawa, T., Uchino, E., Samatsu, T.: The Wavelet Network Using Convex Wavelets and Its Application to Modeling of Dynamical Systems. The Transactions on the IEICE J79-A(12), 2046–2053 (1996) (in Japanese)

Origin of Kanizsa Triangle Illusion

Tomoko Sakiyama, Aisato Sasaki, and Yukio-Pegio Gunji

Dept. of Earth & Planetary Science, Faculty of Science, Kobe University
Rokkodai-cho 1-1, Nada, Kobe 657-8501, Japan
sakiyama@stu.kobe-u.ac.jp, cream923@gmail.com,
yukio@kobe-u.ac.jp

Abstract. In this paper, we introduced Kanizsa triangle figures equipped with stripe notches on its inducers to experimental participants. It is obtained that illusion contour depends on the location and direction of the stripe notches. If notches were on the edge or at inner part of inducers, participants reported less magnitude of the contour illusion. These results suggest that there were plenty of room to consider the mechanism of Kanizsa triangle illusion by paying attention to the information which local inducer carries referring to global functional property.

Keywords: Kanizsa triangle, perception, from part to whole.

1 Introduction

Regarding to perceptions, it's well discussed whether animals, including human, perceive the whole pictures or just element part level [1-4]. However, in any case, it seems that there is a tendency to make judgment which level is true or not, in assuming that each level is independent of each other. Some studies reported that experimental subjects' persistency to either level is changeable depending on situation contexts. [5]. Therefore, there is plenty of space to consider new mechanism to explain of above flexible decisions i.e. employing local element information with referring global property.

Kanizsa triangle figure, in which three Pac-Man are directed inside symmetrically, is well known as illusion that subjective contour is perceived although such contour never exists actually and well referred as good example of gestalt perception [6,7]. In the transcendence perspectives, the apparent triangle appears as a perceptive object based on perceptual processing in visual cortex. Gaetano Kanizsa reported in his own work that these subjective contours might appear from imperfective property which element of this contour (each Pac-man) had [7]. He said that because we had such intrinsic tendency, contour would be formed as prospect objects which could fill the chipped actual objects.

What it is like to perceive the subjective contour which is actual non-existence? Can we explain this without assuming transcendence perspectives?

We assume that perceiving the functional whole as a result which attributes to implication of elements is illusion perception. In other words, we assume that Kanizsa triangle illusion aroused from the two different levels' interaction of part i.e. each

S.-Y. Rhee et al. (eds.), *Soft Computing in Machine Learning*,
Advances in Intelligent Systems and Computing 273,
DOI: 10.1007/978-3-319-05533-6_10, © Springer International Publishing Switzerland 2014

Pac-Man in visual field. Each Pac-Man exists as an object on one hand. On the other hand, however, the global property or meaning of those inducers must be referred as different level. Therefore, introducing the stripe notches which indicate another possibility of parts' global property or meaning, subjective ambiguous might be produced and affect the reliable establishment of contours. Actually, it is reported the possibility of gestalt grouping in unconsciousness [8]. We discuss consistency of our assumption by introducing the perception difference of subjective contour between two visual stimuli which have same imperfective property and tackle to the origin of Kanizsa triangle illusion.

In our working hypothesis, apparent pattern is produced spatially accelerated or inhibited, dependent on a local pattern which serves as bias. We here argue the relationship between the imperfective property mentioned by Kaniza and pattern formation in a bottom up fashion. Especially, we introduced Kanizsa triangle figures equipped with some stripe notches to experimental subjects. To evaluate how much these notches affected establishing relevant clear contour, we investigated whether notches' inserted positions or directions affected contour establishment or not. If implication of inducers (Pac-man) has something ambiguous directional information, positions or directions of stripe notches influence expectation based on local information i.e. inducers. Later, we discuss the consistency between our obtained results and above mentioned assumption.

2 Materials and Methods

Stimulus materials

We made five special Kanizsa triangles. These five figures were named vertical, horizontal (edge), slant, top and hollow, respectively (Fig. 1).

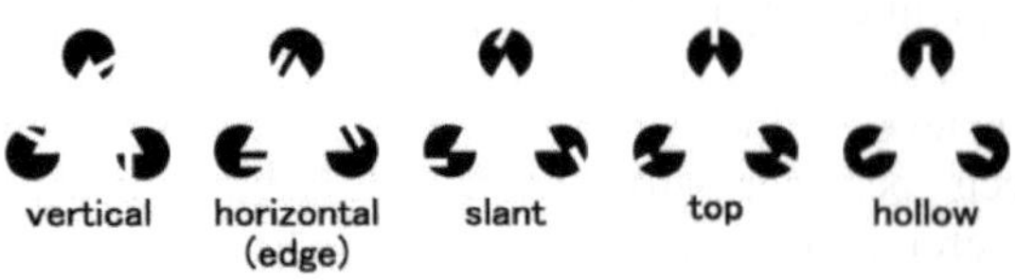

Fig. 1. Five special kanizsa triangles

The diameter of the circles which form triangles were 3.4mm. The distance between any two inducers was 6.8mm. The width of the stripe notches (white bars) were 0.4mm.

After that, We selected two figures from five figures, and made four combinations (Fig. 2).

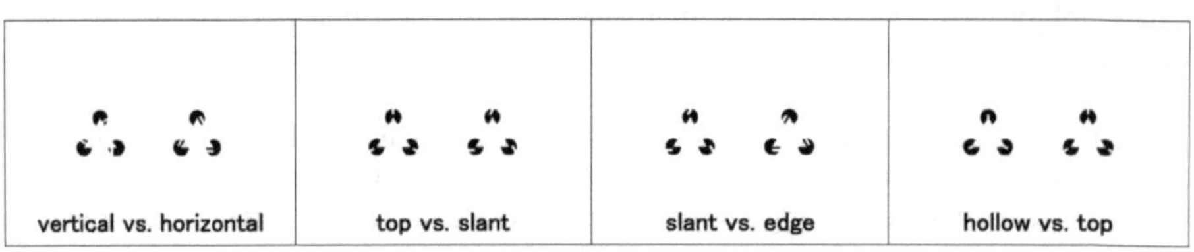

Fig. 2. Four combinations

Here, we made mirrored figures of vertical, horizontal(edge) and slant respectively (Fig. 3).

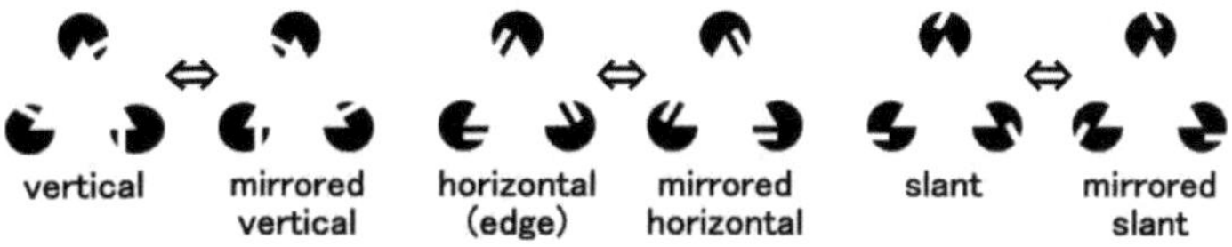

Fig. 3. Mirrored figures

The reason why we made mirrored figures is to check whether there are differences about vision between original triangles and mirrored triangles. Mirrored triangles also made some combinations.

Thus, we had total twelve combinations including mirrored combinations. For example, there are four combinations which are related to vertical and horizontal.

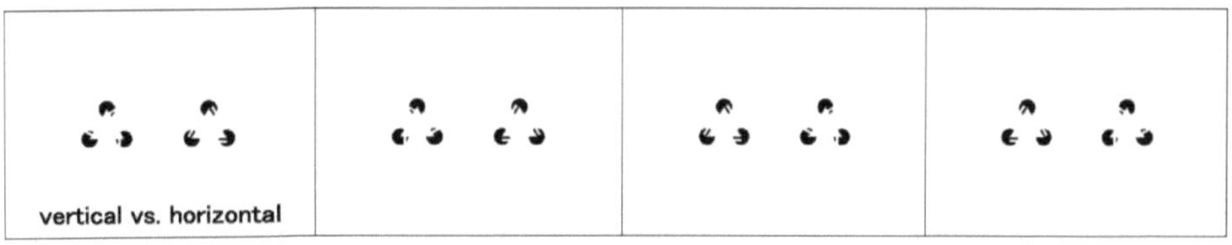

Fig. 4. Four combinations related to *vertical* and *horizontal*

In addition, we prepared four simple Kanizsa triangles (Fig. 5).

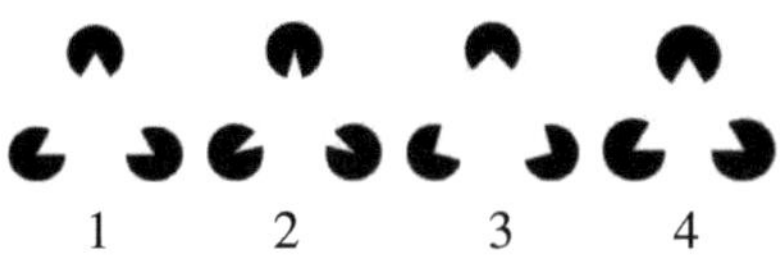

Fig. 5. Four simple Kanizsa triangles

We also selected two figures from four figures, and made eight combinations excepting 2-4, 4-2, 3-4, 4-3 configurations.

So, we have total twenty combinations.

We printed those twenty figures individually on the twenty high-quality cards which were postcard-sized.

Procedure

Participants were fourteen university students. First, they were instructed about procedure of this experiment. After that, they were shown normal Kanizsa triangles and asked whether illusory figures could be seen. If a participant could not see illusion at this stage, then we excluded his or her data. Second, participants were given the twenty cards which were explained above. In addition, they were given the question and answer sheets. Those sheets had two questions per one combination of figures. And two questions were same in all twenty combinations.

Participants turned the twenty cards one by one and answered the questions. Eight combinations of the simple Kanizsa triangles were printed on first eight cards, and twelve combinations of the special Kanizsa triangles followed those. It is because we thought it was very hard for participants to answer the questions about the special Kanizsa triangles without preparations. So we prepared the eight exercises using the simple Kanizsa triangles.

By the way, participants had to have intervals of two seconds between twenty tests to rest their eyes and brains.

Questions

1 st question: intentisity

The first question which was asked about combinations of figures was 'Which figure has the more clear outline of the triangle?' To answer this question, participants were given the following four choices.

1. The left figure has the more clear outline.
2. The right figure has the more clear outline.
3. Both figures have clear outlines. (I can't choose one.)
4. Both figures don't have outlines.

2 nd question: deformation

The second question was 'Do you see a triangle which outline is distorted?' To answer this question, participants were given the four choices again.

1. The left figure has the more distorted triangle.
2. The right figure has the more distorted triangle.
3. Both figures have a distorted triangle. (I can't choose one.)
4. Both figures don't have a distorted triangle.

From the above, the flow of the experiment was following.

1. We instruct about procedure of this experiment.
2. Participants are shown normal Kanizsa triangles and are checked whether they have competence to participate in our experiment.
3. Competent participants are given the twenty cards and the question and answer sheets.
4. Participants turn the card one by one answering the two questions at their own pace, but they have to have two seconds intervals between tests.
5. When the twentieth card is turned and participants finish answering the two questions, this experiment is finished.

3 Results

We formed the results of the above experiment into the four groups. The four groups correspond to four combinations of Fig. 2.

In the combinations of vertical and horizontal, the vertical, horizontal and both rates for the first question were almost same. (Fisher's.test : vertical 19 out of 54, horizontal 17 out of 54 and both 18 out of 54, p=0.946, Fig. 6-1.)
The neither rate for the second question was reasonably high. (Fisher's.test : vertical 1 out of 56, both 1 out of 56 and neither 54 out of 56, p<0.001, Fig. 6-2.)

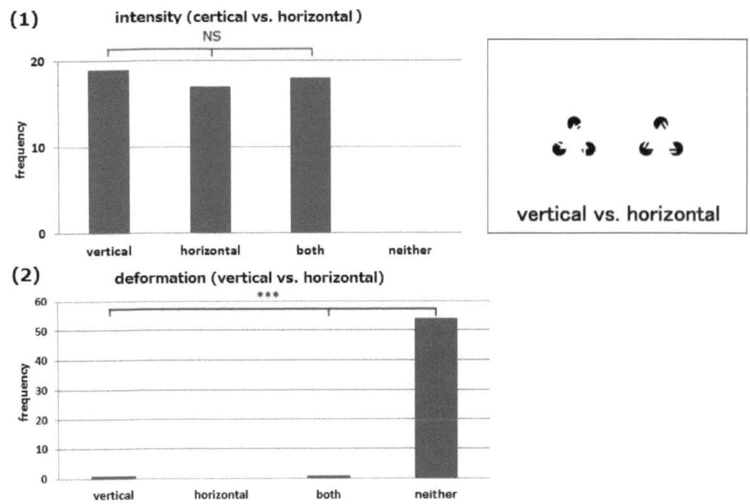

Fig. 6. Experimental results of vertical vs. horizontal (1). intensity. (2) deformation.

In the combinations of top and slant, the top, slant and both rates for the first question were divided (Fisher's.test : top 9 out of 28, slant 5 out of 28 and both 14 out of 28, p=0.1132, Fig. 7-1).

The neither rate for the second question was reasonably high (binomial.test: slant 1 out of 28 and neither 27 out of 28, p<0.001, Fig. 7-2).

In the combinations of slant and edge, participants selected the slant at a high rate in the first question (Fisher's.test : slant 38 out of 56, edge 6 out of 56 and both 12 out of 56, p<0.001, Fig. 8-1).

The neither rate for the second question was reasonably high (Fisher's.test : edge 1 out of 56, both 1 out of 56 and neither 54 out of 56, p<0.001, Fig. 8-2).

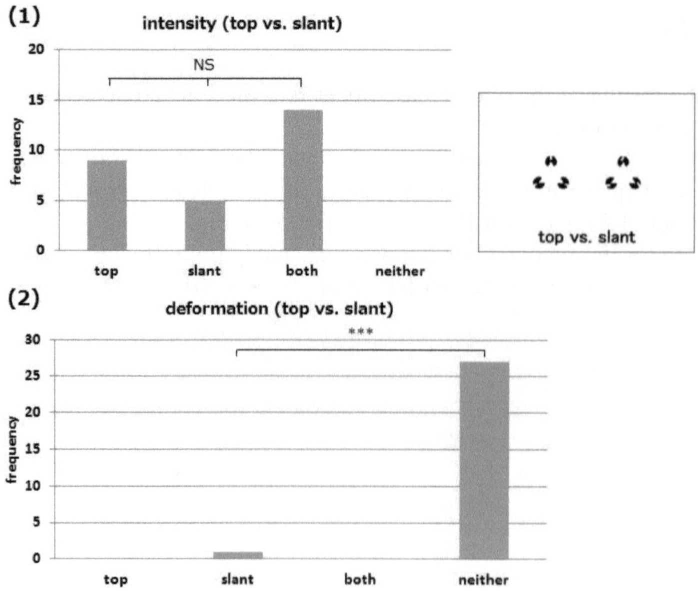

Fig. 7. Experimental results of top vs. slant. (1). intensity. (2) deformation.

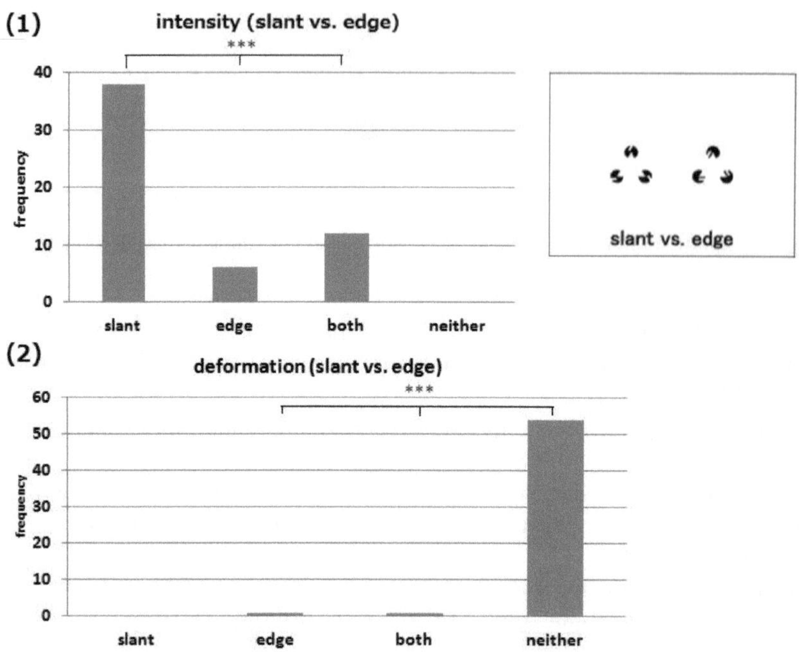

Fig. 8. Experimental results of slant vs. edge. (1). intensity. (2) deformation.

In the combinations of hollow and top condition, participants selected the top at a high rate in the first question (binomial.test: top 25 out of 29 and both 4 out of 29, p<0.001, Fig. 9-1).

And, participants selected the hollow at a high rate in the second question (binomial.test: hollow 20 out of 28 and neither 8 out of 28, p<0.001, Fig. 9-2).

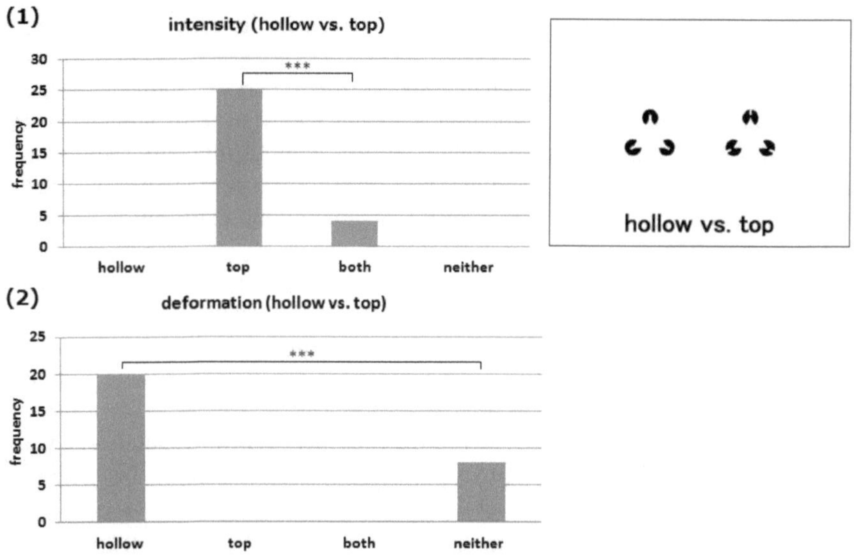

Fig. 9. Experimental results of hollow vs. top. (1). intensity. (2) deformation.

Finally, we checked whether there is consistency in the individual participant's answers among four combinations related to vertical and horizontal (See Fig. 4).

The result is the Fig. 10, and consistency was very high (binomial.test: consistency 43 out of 56 and inconsistency 7 out of 56, p<0.001).

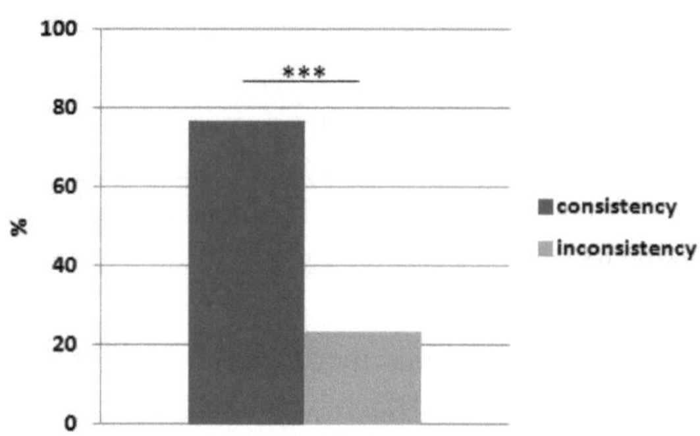

Fig. 10. Consistency of individual participant's answers in vertical vs. horizontal

4 Discussion

Our results show that inserted notches position (whether these are added on top or edge of each Pac-man) influence of subjective contour. It seems that if stripe notches are inserted on edge of inducers, perceived contour intensity is relatively less than that of slant versions. There was no insistency difference between top and slant versions. In addition to that, hollow versions' perceived intensity is less than that of top versions.

Therefore, it might be possible that edge or inner part of inducers relatively contribute to contour perception compared with the Pac-Man with notches in outer part. These illusion figures which have relatively less contour intensity seem to crimp formation of relevant contour between any pair of inducers, perhaps because they let agents to find another possibility as a result of ambiguous exploration from local implicated information.

Our results suggest that local projecting part formed by concave triangle in the Pac-man could be spatially accelerated and connected with each other, which could produce contour perception. Notches near the projecting part (edge, hollow) could disturb producing subjective contour, compared with the Pac-Man with notches in outer part.

We found that there was no intensity difference between horizontal and vertical versions. It is unlikely that dominant eye effects influence that result because of our experimental conditions. Plus, we could obtain that individual perception had consistency regarding vertical and horizontal versions. Therefore, there must be higher-order cognitive processes about that. It's interesting that perceived intensity is different depending on the person. However, it's difficult for us to determine that reason and is needed more investigations.

Agents might perceive illusion contour as objects, through finding or detecting new possibility and stimuli with active expectation which has been triggered by local information passively. Thus, balancing between activeness and passiveness would play an important rule of illusion perception.

5 Conclusion

In this paper, we introduced Kanizsa triangle figures equipped with stripe notches on its inducers to experimental participants and obtained that illusion contour depends on the location and direction of the stripe notches. If notches were on the edge or at inner part of inducers, participants reported less magnitude of the contour illusion.

Acknowledgements. The authors thank the Japan Society for the Promotion of Science for financial support (B2130093).

References

1. Fujita, K.: Perceptual completion in rhesus monkeys (Macaca mulatta) and pigeons (Columba livia). Percep. & Psychophy. 63(1), 115–125 (2001)
2. Hopkins, W.D., Washburn, D.A.: Matching visual stimuli on the basis of global and local features by chimpanzees (Pan troglodytes) and rhesus monkeys (Macaca mulatta). Anim. Cogn. 5, 27–32 (2002)
3. Nagasaka, Y., Hori, K., Osada, Y.: Perceptual grouping in pigeons 32, 625–632 (2005)
4. Vallortigara, G., Snyder, A., Kaplan, G., Bateson, P., Clayton, N.S., Rogers, L.J.: Are Animals Autistic Savants? PLOS Biol. 6(2), 0208–0214 (2008)
5. Matsukawa, A., Inoue, S., Jitsumorivol, M.: Pigeon's recognition of cartoons: effects of fragmentation, scrambling, and deletion of elements. Behav. Proc. 65, 25–34 (2004)
6. Kanizsa, G.: Subjective contours. Sci. Am. 234, 48–52 (1976)
7. Kanizsa, G.: Organisation in Vision. Praeger, New York (1979)
8. Wang, L., Weng, X., He, S.: Perceptual Grouping without Awareness: Superiority of Kanizsa Triangle in Breaking Interocular Suppression. PLOS ONE 7(6), 1–6 (2012)

LTS-SVMR for Modeling of Nonlinear Systems with Noise and Outliers

Chen-Chia Chuang[1], Jin-Tsong Jeng[2,*], and Guan-Yi Hu[2]

[1] Department of Electrical Engineering,
National Ilan University, Ilan County, Taiwan ROC
[2] Department of Computer Science and Information Engineering,
National Formosa University, Yunlin County, Taiwan ROC
tsong@nfu.edu.tw

Abstract. In general, there are many machine learning algorithms have developed for the intelligent control. Besides, support vector machine (SVM) among machine learning is generally used in recent years. That is, many contributions about the support vector machine regression (SVMR) and the least squares-support vector machine regression (LS-SVMR) can be found in some well-known journals. In this paper, for the robustness problem of LS-SVMR, we propose the least trimmed squares support vector machine regression (LTS-SVMR) which is the hybrid of the least trimmed squares (LTS). That is, when the LTS method faces on the training sample with noise and outliers, it can effectively remove large noise and outliers under the proper initial nonlinear function. That is, robustness of the LS-SVMR is enhanced by combining the LS-SVMR and the LTS. Finally, the proposed LTS-SVMR is applied on modeling of nonlinear systems with noise and outliers.

Keywords: Least squares support vector machine regression, least trimmed squares, modeling, noise and outliers.

1 Introduction

Support vector machine (SVM) is a supervised learning algorithm in machine learning and was proposed by Vapnik in 1998 [1]. A basic concept of SVM is to construct the optimal hyperplane in the high-dimensional space which comes from a kernel function mapping results. Besides, the optimal hyperplane has a maximum margin. In recent, some papers [2-5] that can be found for new applications and improvements about SVM. Two major category of the SVM are classification and regression. That is, the SVM is classified into the support vector machine regression (SVMR) and the support vector machine classification (SVMC). In this paper, we focus on the regression analysis. The SVMR is a quadratic programming (QP) problem. In the SVMR, the optimal hyperplane is obtained by solving the QP problem and is used to approximate the training data set and we can use it to do prediction. Besides, support vectors

* Corresponding author.

S.-Y. Rhee et al. (eds.), *Soft Computing in Machine Learning*,
Advances in Intelligent Systems and Computing 273,
DOI: 10.1007/978-3-319-05533-6_11,

play important roles that control a state of the hyperplane and avoid the overfittig phenomenon in the SVMR.

The LS-SVMR [6-8] is the improved version of SVMR on learning speed. The major problem about the SVMR is to find the optimal hyper plane which is also the optimal approximate function and this is as the convex optimization problem. The sparse solution of a convex optimization problem is obtained by solving a QP problem. However, a QP problem leads the formulation SVMR to become more complex. In order to overcome this problem, Suykens proposed the LS-SVMR method in 1999 years. The concept of this method is that inequality constraints of the primal optimization problem of the SVMR modify an equation and an ε-loss function of the SVMR is replaced by a regularized least squares cost function. That is, LS-SVMR is based on this concept makes a complex QP problem transform into a simple linear equation problem [6]. However, this improvement causes the robustness problem on LS-SVMR. That is, the LS-SVMR using a regularized least squares cost function has the major drawback which its modeling is balefully influenced by training sample with outlier observations and this fact is indicated by those literatures [2, 11-13]. Hence, the SVMR with weighted least square (WLS-SVMR) in [6, 7, 12, 14] was proposed for overcoming the robustness problem of the LS-SVMR. The WLS-SVMR procedure is constructed by main three steps. First step is to once do the estimated LS-SVMR for the training sample. In second step, weighting values are calculated by the hample weight function with error variables and a robust scale estimator. Finally, a new model based on weighting values is reconstructed.

In general, for modeling of the training data contains outliers or noises; outlier may be an experimental error of to be artificial or the unique character of a data and occurs in the tail portion of some noise distributions. Noise occurs in some nature phenomenon. According to [9] the outlier can be classified into three types is mentioned. There are good leverage points, bad leverage points and vertical outliers; bad leverage points and vertical outliers are danger for modeling. In this paper, we propose that a new method of enhancing robustness of the LS-SVMR is the hybrid of the least trimmed squares (LTS) algorithm and the LS-SVMR. That is, two LTS-SVMRs with different initial model are proposed. The LTS method that is proposed via Rousseeuw [10] is the statistical technique which is able to provide robust regression and remove outlying observations. It uses a minimized sum of the squared residuals to estimate a clear sample and then do modeling of a clear sample based on least square (LS) estimator. Breakdown points of the LTS is 50% means an ability of modeling of it keep a correct state in that half data are vertical outliers or bad leverage points. In this paper, the proposed LTS-SVMR is used in modeling of nonlinear dynamic systems with noise and outliers. That is, proposed the new LTS-SVMR method which is the hybrid of the LTS and the LS-SVMR is employed in modeling a nonlinear MIMO system for the sample with outliers and noises. The architecture of that modeling of nonlinear MIMO systems using the LTS-SVMR contains two stages. Besides, noises and outliers that are independently generated by Gaussian distribution, Cauchy distribution and student's T distribution add to train data of the nonlinear MIMO systems for the proposed LTS-SVMR.

There are remaining sections are outlined follows in this paper and are as follows. In section 2, the SVMR, the LS-SVMR, the WLS-SVMR and the LTS are brief reviewed and the proposed LTS-SVMR with robustness and its application which is modeling of chaotic time series systems are introduced. Section 3 shows simulation results based on the LTS-SVMR. Section 4 summarizes a conclusion according to those results and characters of the proposed LTS-SVMR.

2 Research Methods

2.1 Least Squares Support Vector Machines Regression

In LS-SVMR [6], the complex QP problem based on that a regularized least squares cost function replaces an ε loss function is transformed into a linear system problem. A regularized least squares cost function is defined as

$$\min \frac{1}{2}\sum_{i=1}^{n} e_i^2,\ e_i = y_i - (w^T \phi(x_i) + b),\ i = 1,\ldots,n\,. \tag{1}$$

Error variables e_i of Eq. (1) are similar to slack variables ξ_i in the SVMR formulation. It can allow that error between predicting values and true values. Besides, each training data point has different value of error variable. The primal optimization problem about the LS-SVMR is represented as

$$\begin{aligned} \min\quad & \frac{1}{2} w^T w + \gamma \frac{1}{2}\sum_{i=1}^{n} e_i^2, \\ s.t.\quad & y_i = w^T \phi(x_i) + b + e_i,\quad i = 1,\ldots,n, \end{aligned} \tag{2}$$

where γ is a regularization parameter and we can avoid the over fitting problem by adjusting its size. Besides, from Eq. (2) we can know that is primal optimization problem with constraints. In order to transform into a dual problem, Lagrangian for the primal optimization problem is defined as

$$L = \frac{1}{2} w^T w + \gamma \frac{1}{2}\sum_{i=1}^{n} e_i^2 - \sum_{i=1}^{n} \alpha_i (w^T \phi(x_i) + b + e_i - y_i), \tag{3}$$

where α_i are the Lagrange multipliers. Four conditions are generated by doing a partial differential of Eq. (3) with respect to w, b, e, α and there are

$$\begin{cases} \dfrac{\partial L}{\partial w} = 0 & \rightarrow \quad w = \sum_{i=1}^{n} \alpha_i \phi(x_i), \\ \dfrac{\partial L}{\partial b} = 0 & \rightarrow \quad \sum_{i=1}^{n} \alpha_i = 0, \\ \dfrac{\partial L}{\partial e_i} = 0 & \rightarrow \quad \alpha_i = \gamma e_i, \\ \dfrac{\partial L}{\partial \alpha_i} = 0 & \rightarrow \quad w^T \phi(\mathrm{x}_i) + \mathrm{b} + \mathrm{e}_i - \mathrm{y}_i = 0. \end{cases} \tag{4}$$

From Eq. (4) we can infer a linear system which is represented as

$$\begin{bmatrix} 0 & 1_v^T \\ 1_v & \Omega + I/\gamma \end{bmatrix}_{(n+1)\times(n+1)} \begin{bmatrix} b \\ \alpha \end{bmatrix}_{(n+1)\times 1} = \begin{bmatrix} 0 \\ Y \end{bmatrix}_{(n+1)\times 1}, \tag{5}$$

where $Y = [y_1;\ldots;y_n]$, $1_v = [1;\ldots;1]$, $\alpha = [\alpha_1;\ldots;\alpha_n]$ and $\Omega = \phi(x)^T \phi(x_i)$, $i = 1,\ldots,n$. In the LS-SVMR we applies a kernel function $K(x, x_i)$ to replace $\Omega = \phi(x)^T \phi(x_i)$. Hence, the nonlinear regression function of the LS-SVMR is modified as

$$f(x) = \sum_{i=1}^{n} \alpha_i K(x, x_i) + b, \tag{6}$$

where solutions α_i, b about are obtained by solving Eq. (5).

2.2 The Least Trimmed Square

The basic concept about the LTS method [15-16] can be described. We consider the training data and the LTS objective function is defined as

$$\min \sum_{i=1}^{h} (r_i^2)_{i:n}, \tag{7}$$

where r_i is an error between a predicted value and a true value, $\left(r_i^2\right)_{i:n}$ represents the i-th location of the queue which is that square residuals are orderly sorted, i.e. $(r_1^2)_{1:n} \le (r_2^2)_{2:n} \le \cdots \le (r_{n-1}^2)_{n-1:n} \le (r_n^2)_{n:n}$, h is trimming constant and its value may be between n and n/2. Eq. (7) represents that the special subsample which makes the total sum of its square residuals to be minimized and is found from all subsample of size h. Moreover, from Eq. (7) we can know a large value of a square residual of $n-h$ observations which implies that those observations may be vertical outliers or bad leverage points. In other words, contaminated observations can be detected and removed by the LTS method, but that breakdown point of the LTS is zero means that the LTS method becomes the LS method when trimming constant h equals to n.

2.3 The Least Trimmed Square Support Vector Machine Regression for Modeling of Nonlinear Systems with Noise and Outliers

In this paper, the proposed LTS-SVMR improves the low robustness of the LS-SVMR which is due to that it adopts a sum squared error cost function to replace a ε loss function so it must consider outliers in the parameters estimation process. The reason for the hybrid of the LS-SVMR and the LTS is that the LTS has the ability of removing outliers and maximal 50% breakdown point which means the LTS estimator still keep a good robustness when a half number of the training sample belongs to a outlier point. Therefore, we expect that the LS-SVMR based on the ideal of the LTS can handle with the training sample which contains outliers. The LTS-SVMR is used in modeling of the training sample. The nonlinear regression function $f(x)$ based on the training sample is obtained by the LS-SVMR estimator. Then, to compute squared residuals r_i is as follows:

$$r_i = (y_i - f(x_i))^2, \tag{8}$$

where x_i is a input data, y_i is a target data. Those computed squared residuals need to be sorted and a sort result of those is the form:

$$(r_1^2)_{1:n} \le (r_2^2)_{2:n} \le \cdots \le (r_{n-1}^2)_{n-1:n} \le (r_n^2)_{n:n} \,.$$

To assume existing optimal trimming constant h has to be between n/2 and n. That indicates that exiting n-h training data with largest squared residuals may to be outliers and leads the LTS-SVMR estimator to have a good robustness. And, the LTS-SVMR estimator is defined as

$$(w^{'},b^{'})^{LTS} = \arg\min \sum_{i=1}^{h}(y_i - g(x_i))^2 = \arg\min \sum_{i=1}^{h}(y_i - w^{'T}\phi(x_i)+b^{'})^2, \tag{9}$$

where $w^{'} \in \Re^{p}$ is the unknown weight vector, $b^{'} \in \Re$ is the unknown bias and $g(x_i) = w^{'T}\phi(x_i) + b^{'}$ is the nonlinear regression function of the LTS-SVMR. For the LTS-SVMR estimator, the optimization problem is represented as

$$\begin{aligned} &\min \quad \frac{1}{2}w^{T}w + \gamma\frac{1}{2}\sum_{i=1}^{h} e_i^{'2}, \\ &s.t. \quad y_i = w^{'T}\phi(x_i) + b^{'} + e_i^{'}, \quad i = 1,\ldots,h, \end{aligned} \tag{10}$$

where $e_i^{'}$ are error variables. This optimization problem use Lagrange to transform into the dual problem and its Lagrange is as follows:

$$L = \frac{1}{2}w^{'T}w^{'} + \gamma\frac{1}{2}\sum_{i=1}^{h} e_i^{'2} - \sum_{i=1}^{h}\alpha_i^{'}(w^{'T}\phi(x_i) + b^{'} + e_i^{'} - y_i), \tag{11}$$

where $\alpha_i^{'}$ are the Lagrange multipliers. After doing partial differential of Eq. (11) with respect to $w^{'}$, $b^{'}$, $e^{'}$, $\alpha^{'}$ and we can obtain Eq. (12) which is

$$\begin{cases} \dfrac{\partial L}{\partial w^{'}} = 0 \;\rightarrow\; w^{'} = \sum_{i=1}^{h}\alpha_i^{'}\phi(x_i), \\ \dfrac{\partial L}{\partial b^{'}} = 0 \;\rightarrow\; \sum_{i=1}^{h}\alpha_i^{'} = 0, \\ \dfrac{\partial L}{\partial e_i^{'}} = 0 \;\rightarrow\; \alpha_i^{'} = \gamma e_i^{'}, \\ \dfrac{\partial L}{\partial \alpha_i^{'}} = 0 \;\rightarrow\; w^{'T}\phi(x_i)+b^{'}+e_i^{'}-y_i=0. \end{cases} \tag{12}$$

From Eq. (12) we can infer a linear system which is represented as

$$\begin{bmatrix} 0 & 1_v^T \\ 1_v & \Omega_h + I/\gamma \end{bmatrix}_{(h+1)\times(h+1)} \begin{bmatrix} b' \\ \alpha' \end{bmatrix}_{(h+1)\times 1} = \begin{bmatrix} 0 \\ Y_h \end{bmatrix}_{(h+1)\times 1}, \tag{13}$$

where $Y_h = [y_1;\ldots;y_h]$, $1_v = [1;\ldots;1]$, $\alpha' = [\alpha_1';\ldots;\alpha_h']$ and $\Omega = \phi(x)^T \phi(x_i), i = 1,\ldots,h$. In the LTS-SVMR we applies a kernel function $K(x_i, x_j)$ to replace $\Omega = \phi(x)^T \phi(x_i)$. Therefore, the nonlinear regression function of LTS-SVMR is modified as

$$f(x) = \sum_{i=1}^{h} \alpha_i' K(x, x_i) + b', \tag{14}$$

where solutions of α_i', b' are obtained by solving a linear system in Eq. (13) with the RBF kernel. The proposed algorithm is follows that:

(1) For inputting training sample, an initial model is obtained by estimate of once LS-SVMR or random selection.
(2) For training sample, prediction is done by an initial model. And then, squared residuals r_i^2 between target values and predicted values are computed. In order to get the ordered squared residuals, they are sorted by the rule which is follows as $r_1^2 \le r_2^2 \le \cdots \le r_{n-1}^2 \le r_n^2$.
(3) The trimming process is done and is that several data based on several large squared residuals in the ordered squared residuals are ignored (trimmed) and other data makes a new subsample. Besides, the next number of trimming data is that the now number of trimming data pluses a fixed constant. A new model adopts the LS-SVMR to be built for a new subsample and Prediction based on a new model is performed.
(4) We chooses a good model from those new models when a number of loop don't equal to the half number of the training sample, otherwise it has to back the step 2.

In general, the process industries can't use a simple mathematical function to be represented because it must consider many uncertain factors, numerous variables and existing disturbances [2]. Hence, nonlinear MIMO systems are more difficulty for modeling. For this problem, there are many approaches [5-7] that proposed. In this paper, we proposed the new LTS-SVMR method which is the hybrid of the LTS and the LS-SVMR is employed for modeling a nonlinear MIMO system under the sample with outliers and noises. The architecture of that modeling of nonlinear MIMO systems using the LTS-SVMR is shown in Fig. 1 and generally contains two stages. The noises and outliers for the proposed approach are independently generated by Gaussian distribution, Cauchy distribution or student's T distribution to training data of the nonlinear MIMO systems. In first stage, phase spaces that are training samples of the LTS-SVMR use a delay embedding theorem of Takens' theorem [13] to be reconstructed for training data. The dynamic system is calculated by the form:

$$x_{i+1+(m-1)\tau} = f(x_i, x_{i+\tau}, \ldots, x_{i+(m-1)\tau}), \tag{15}$$

where the embedding variable m sets seven and the time delay variable τ sets three, $x_i, x_{i+\tau}, \ldots, x_{i+(m-1)\tau}$ is input data and $x_{i+1+(m-1)\tau}$ is output data (target). The dynamic learning system is calculated by the form:

$$\tilde{x}_{i+1+(m-1)\tau} = f(x_i, x_{i+\tau}, \ldots, x_{i+(m-1)\tau}), \tag{16}$$

According to the proposed LTS-SVMR approaches, there are three different types noise for the modeling of nonlinear MIMO systems using the LTS-SVMR in stage 1. In second stage, the model based on the LTS-SVMR is built for the dynamic system and the LTS-SVMR uses the RBF kernel. Besides, the LTS-SVMR method has two different types; namely, using once LS-SVMR and random selected data point to determine the initial model. After finishing the procedure of Fig. 1, the prediction process is done by this appropriate training model for a testing data.

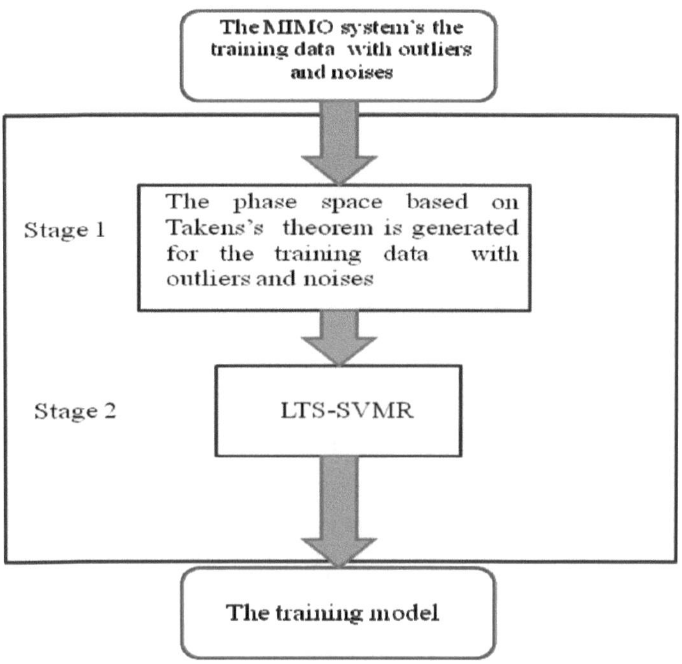

Fig. 1. The procedure of the modeling of nonlinear MIMO systems with outliers and noises uses the proposed LTS-SVMR

3 Simultion Results

In this section, the proposed LTS-SVMR method is to enhance a robustness of the LS-SVMR and two types for initial model on LTS-SVMR is used. This section shows simulation results of the new LTS-SVMR method for the modeling of nonlinear dynamic systems. The simulations can be classified into three parts according to that using type of noises and outliers. Noises and outliers can be generated by Cauchy, normal and student's t distributions. Table 1 shows the parameters of the noises in this paper.

Table 1. The parameters of the noise distributions

Noise	P.D.F.	Parameter
Cauchy	$\frac{1}{\pi}\left(\frac{\gamma}{(x-x_0)^2+r^2}\right)$	$\gamma = 1$
Normal	$\frac{1}{\sigma\sqrt{2\pi}}\exp(-\frac{(x-\mu)^2}{2\sigma^2})$	$\mu = 0$ $\sigma = 0.1$
Student's t	$\frac{\Gamma(\frac{v+1}{2})}{\sqrt{v\pi}\Gamma(\frac{v}{2})(1+\frac{x^2}{v})^{\frac{v+1}{2}}}$	$v=3$

The measure of simulation results is to use the root mean square error (RMSE) which is defined as

$$RMSE = \sqrt{\frac{\sum_{i=1}^{N}(y_i - \hat{y}_i)^2}{N}} \tag{17}$$

where y_i is the i-th target value, $\hat{y}_i$ is the i-th predicted value and N is the number of data. The simulation for modeling of nonlinear MIMO system to be identified is defines as [17]

$$\begin{bmatrix} y_1(k+1) \\ y_2(k+1) \end{bmatrix} = \begin{bmatrix} \frac{y_1(k)}{1+y_2^2(k)} \\ \frac{y_1(k)y_2(k)}{1+y_2^2(k)} \end{bmatrix} + \begin{bmatrix} x_1(k) \\ x_2(k) \end{bmatrix}, \tag{18}$$

where $(x_1(k), x_2(k)) = (\sin(\frac{2\pi k}{25}), \cos(\frac{2\pi k}{25}))$, $0 \le k \le 1500$. In this paper, the output y1 and y2 of nonlinear dynamic systems generates two types of the sample with 1501 data points and three types of the noises (i.e. Cauchy, normal and student's t) are added into those data. 1501 data points with the noises are employed to do the

Table 2. Shown the compared RMSE of simulation results for the modeling output y1of the nonlinear MIMO systems using the RBF kernel

Method \ Noise		Cauchy	Normal	Student's t
SVMR	Training	4.21	0.34	2.43
	Testing	2.28	0.57	1.9
LS-SVMR	Training	1.25	0.25	1.21
	Testing	1.34	0.26	1.31
WLS-SVMR	Training	0.75	0.26	1.16
	Testing	1.04	0.28	1.28
LTS-SVMR (the choosing the initial model with once LS-SVMR estimation)	Training	0.69	0.2	1
	Testing	0.53	0.19	1
LTS-SVMR (the choosing the initial model with random selection)	Training	0.64	0.2	0.98
	Testing	0.52	0.19	0.98

Table 3. Shown the compared RMSE of simulation results for the modeling output y2 of the nonlinear MIMO systems using the RBF kernel

Method \ Noise		Cauchy	Normal	Student's t
SVMR	Training	3.96	0.97	2.48
	Testing	1.41	0.97	1.07
LS-SVMR	Training	3.26	0.74	1.51
	Testing	1.46	0.79	1.05
WLS-SVMR	Training	0.69	0.74	1.28
	Testing	0.83	0.79	1.03
LTS-SVMR (the choosing the initial model with once LS-SVMR estimation)	Training	0.6	0.62	0.9
	Testing	0.62	0.6	0.87
LTS-SVMR (the choosing the initial model with random selection)	Training	0.61	0.62	0.86
	Testing	0.59	0.59	0.86

phase space reconstruction process. Then, the new sample with 1000 data points that can constructs the phase space of a output of nonlinear dynamic systems is obtained. For building a nonlinear regression function of the nonlinear MIMO systems, the training sample is first 500 data points of the new sample and the testing data is last 500 data points of the new sample. Finally, all regression methods (i.e. SVMR, LS-SVMR, the proposed LTS-SVMR methods) based on those samples are executed. Finally, the modeling of nonlinear MIMO systems with noise and outliers has compared with SVMR, LS-SVMR and WLS-SVMR and the proposed LTS-SVMR. The results in Tables 2 and 3 show the compared results based on the modeling method in this paper.

4 Conclusions

In this paper, there are five machine learning algorithms that are compared for modeling of nonlinear MIMO systems with noise and outliers. The major purpose of the proposed LTS-SVMR is to overcome complex computation of the SVM that is caused by solving a QP problem using a regularized least squares cost function which replaces ε loss function. At the same time, the LTS-SVMR can handle the training data with noise and outliers for modeling and overcome that easily occurs overfitting in LS-SVMR. That is, we proposed the LTS-SVMR which is the hybrid of the LS-SVMR and the LTS to enhance the robustness of the LS-SVMR and applied it to do modeling of the nonlinear dynamic systems. According to the simulation results, the proposed LTS-SVMR can effectively prevent danger influence of noise and outliers for the modeling process.

Acknowlegdements. The authors wish to thank that this work was supported by National Science Council Under Grant NSC 100-2221-E-150-045, NSC 101-2221-E-150-048-MY2 and National Formosa University EN99B-B1005.

References

[1] Karatzoglou, A., Meyer, D.: Support Vector Machines in R. Journal of Statistical Software 15(9), 1–28 (2006)

[2] Chuang, C.C., Lai, M.H., Chen, S.S., Jeng, J.T.: Hybrid robust LS-SVMR with outliers for MIMO system. In: IEEE International Conference on Systems Man and Cybernetics (SMC), pp. 10–13 (October 2010)

[3] Quan, T., Liu, X., Liu, Q.: Weighted Lest Squares Support Vector Machine Local Region Method for Nonlinear Time Series Prediction. Applied Soft Computing 10(2), 562–566 (2010)

[4] Li, Z., Tang, X.: Using Support Vector Machines to Enhance the Perform-ance of Bayesian Face Recognition. IEEE Transactions on Information Forensics and Security 2(2), 174–180 (2007)

[5] Camps-Valls, G., Soria-Olivas, E., Perez-Ruixo, J.J., Perez-Cruz, F., Artes-Rodriguez, A., Jimenez-Torres, N.V.: Therapeutic Drug Monitoring of Kidney Transplant Recipients Using Profiled Support Vector Machines. IEEE Transactions on Systems, Man, and Cybernetics, Part C: Aplications and Reviewa 37(3), 359–372 (2007)

[6] Suykens, J.A.K., Gestel, T.V., Brabanter, J.D., Moor, B.D., Vandewalle, J.: Least Squares Support Vector Machines, ch. 2. World Scientific, New Jersey (2002)
[7] Valyon, J., Horvath, G.: A Weighted Generalized LS–SVM. Periodica Polytechnica Ser. El. Eng. 47(3), 229–251 (2003)
[8] Yu, Z., Cai, Y.: Least Squares Wavelet Support Vector Machines for Nonlinear System Identification. In: Wang, J., Liao, X.-F., Yi, Z. (eds.) ISNN 2005. LNCS, vol. 3497, pp. 436–441. Springer, Heidelberg (2005)
[9] Rousseeuw, P.J., Leroy, A.M.: Robust Regression and outlier Detection. John Wiley & Sons, Inc., Hoboken (2003)
[10] Rousseeuw, P.J.: Multivariate Estimation With High Breakdown Point. In: Grossmann, W., Pflug, G., Vincze, I., Werty, W. (eds.) Mathematical Statistics and Applications, vol. B, pp. 283–297. Reidel, Dordrecht (1985)
[11] Alpaydm, E.: Introduction to Machine Learning, ch.1. The MIT Press (October 2004)
[12] Quan, T., Liu, X., Liu, Q.: Weighted Lest Squares Support Vector Machine Local Region Method for Nonlinear Time Series Prediction. Applied Soft Computing 10(2), 562–566 (2010)
[13] Chuang, C.C., Su, S.F., Jeng, J.T., Hsiao, C.C.: Robust Support Vector Regression Networks for Function Approximation With Outliers. IEEE Transactions on Neural Networks 13(6), 1322–1330 (2002)
[14] Suykens, J.A.K., De Brabanter, J., Lukas, L., Vandewalle, J.: Weighted least squares support vector machines: robustness and sparse approximation. Neurocomputing 48(1-4), 85–105 (2002)
[15] Rousseeuw, P.J., Driessen, K.V.: Computing LTS Regression for Large Data Sets. Data Mining and Knowledge Discovery 21(1), 29–45 (2006)
[16] Leontitsis, A., Pange, J.: Statistical significance of the LMS regression. Mathematics and Computers in Simulation 64(5), 543–547 (2004)
[17] Fu, Y.Y., Wu, C.J., Jeng, J.T., Ko, C.N.: Identification of MIMO systems using radial basis function networks with hybrid learning algorithm. Applied Mathematics and Computation 213(1), 184–196 (2009)

Author Index

MIX
Papier aus verantwortungsvollen Quellen
Paper from responsible sources
FSC® C105338

If you have any concerns about our products,
you can contact us on
ProductSafety@springernature.com

In case Publisher is established outside the EU,
the EU authorized representative is:
Springer Nature Customer Service Center GmbH
Europaplatz 3, 69115 Heidelberg, Germany

Printed by Libri Plureos GmbH
in Hamburg, Germany